Organic Chemistry

Online Diagnostic Test

Go to **Schaums.com** to launch the Schaum's Diagnostic Test.

This convenient application provides a 30-question multiple-choice test that will pinpoint areas of strength and weakness to help you focus your study. Questions cover all aspects of organic chemistry, and the correct answers are explained in full. With a question-bank that rotates daily, the Schaum's Online Test also allows you to check your progress and readiness for final exams.

Other titles featured in Schaum's Online Diagnostic Test:

Schaum's Easy Outlines: Calculus, 2nd Edition
Schaum's Easy Outlines: Geometry, 2nd Edition
Schaum's Easy Outlines: Statistics, 2nd Edition
Schaum's Easy Outlines: Elementary Algebra, 2nd Edition
Schaum's Easy Outlines: College Algebra, 2nd Edition
Schaum's Easy Outlines: Biology, 2nd Edition
Schaum's Easy Outlines: Human Anatomy and Physiology, 2nd Edition
Schaum's Easy Outlines: Beginning Chemistry, 2nd Edition
Schaum's Easy Outlines: College Chemistry, 2nd Edition

SCHAUM'S™ EASY OUTLINES

Organic Chemistry

Second Edition

Herbert Meislich, Howard Nechamkin, and Jacob Sharefkin

Abridgement Editors:
Mark S. Meier and Jennifer L. Muzyka

New York Chicago San Francisco Lisbon London Madrid Mexico City
Milan New Delhi San Juan Seoul Singapore Sydney Toronto

The ***McGraw-Hill*** *Companies*

1 2 3 4 5 6 7 8 9 10 11 12 13 14 15 WFR/WFR 1 9 8 7 6 5 4 3 2 1 0

ISBN 978-0-07-174590-1
MHID 0-07-174590-4

Library of Congress Cataloging-in-Publication Data

Moyer, Robert E.
Meislich, Herbert.
Schaums easy outline of organic chemistry / Herbert Meislich. — 2nd ed.
p. cm. — (Schaum's outlines)
Includes index.
ISBN 0-07-174590-4 (alk. paper)
1. Chemistry, Organic—Outlines, syllabi, etc. 2. Chemistry, Organic—Problems, exercises, etc. I. Title II. Title: Organic chemistry.

QD256.5.M45 2010
547—dc22 2010010891

This book is printed on acid-free paper.

Contents

Chapter 1
STRUCTURE AND PROPERTIES

IN THIS CHAPTER:

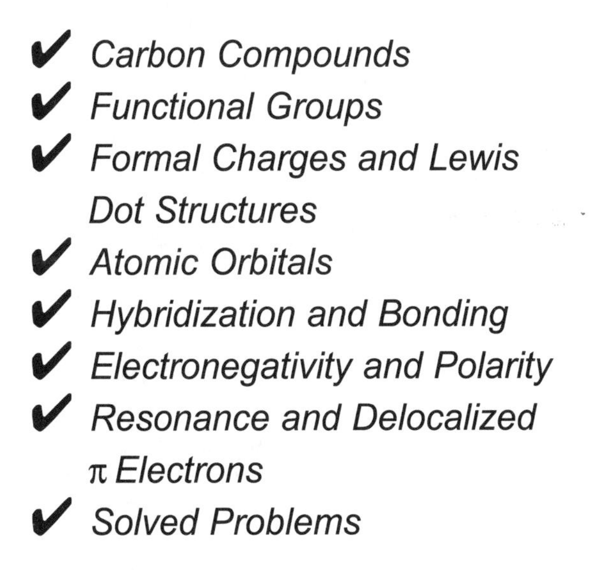

Carbon Compounds

Organic chemistry is the study of carbon (C) compounds, molecules which have **covalent** bonds. Carbon atoms can bond to each other to form open-chain compounds, or **cyclic (ring)** compounds. Both types can also have branches of C atoms. **Saturated** compounds have C's bonded to each other by **single** bonds, **C—C; unsaturated** compounds have C's joined by **multiple** bonds. Examples with **double** bonds and **triple** bonds are shown below. Cyclic compounds having at least one atom in the ring other than C (a **heteroatom)** are called **heterocyclics**. The heteroatoms are usually oxygen (O), nitrogen (N), or sulfur (S).

Most carbon-containing molecules are three-dimensional. In methane, the bonds of C make equal angles of 109.5° with each other, and each of the four H's is at a vertex of a regular tetrahedron whose center is occupied by the C atom. Other shapes do occur: ethene, for example, is planar, and ethyne (acetylene) is linear.

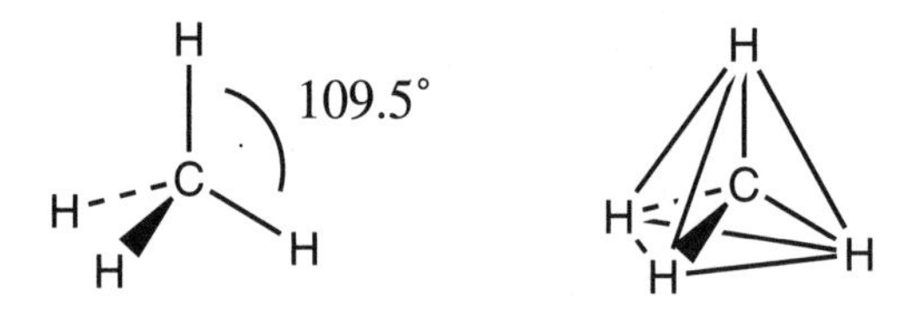

Methane is a tetrahedron

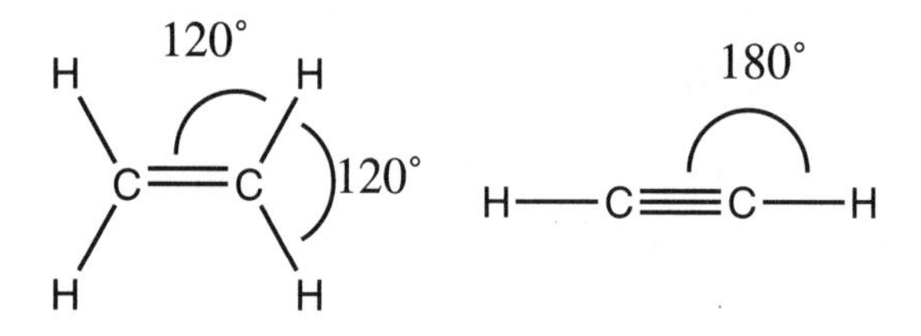

Ethene Ethyne

Organic compounds show a widespread occurrence of **isomers,** which are compounds having the same molecular formula but different structural formulas. Isomers have different chemical and physical properties. This phenomenon of **isomerism** is exemplified by isobutane and n-butane. The number of isomers increases as the number of atoms in the molecule increases.

$$CH_3CH_2CH_2CH_3$$

n-Butane

$$\begin{array}{c} CH_3 \\ | \\ CH_3CHCH_3 \end{array}$$

Isobutane

Functional Groups

Hydrocarbons contain only C and hydrogen (H). H's in hydrocarbons can be replaced by other atoms or groups of atoms. These replacements, called **functional groups,** are the reactive sites in molecules. Double and triple bonds are considered to be functional groups. Some common functional groups are given in the Functional Group Table. The "R" group is a generic group, and is not part of the functional group of interest. Compounds with the same functional group form a **homologous series** having similar chemical properties and often exhibiting a regular gradation in physical properties with increasing molecular weight.

Formal Charge

The formal charge on a covalently bonded atom equals the number of valence electrons of the unbonded atom minus the number of electrons assigned to the atom in its bonded state. The assigned number is one half the number of shared electrons plus the total number of unshared electrons. The sum of all formal charges in a molecule equals the charge on the species. In this outline formal charges and actual ionic charges are both indicated by the signs + and – . The structures shown below are called **Lewis dot structures** (or simply dot structures). Each dot represents an electron in the outer shell of the atom. These drawings can be highly useful in determining if an atom bears a formal charge.

·N̈· (5 dots around N)	H:N̈:H over H	[H:N̈:H, with H above and H below]$^{+}$
Neutral N atom (5 e^-)	Neutral NH_3 (5 e^- from N, 3 e^- from H's)	Ammonium cation (8 total e^-, 4 belong to N)

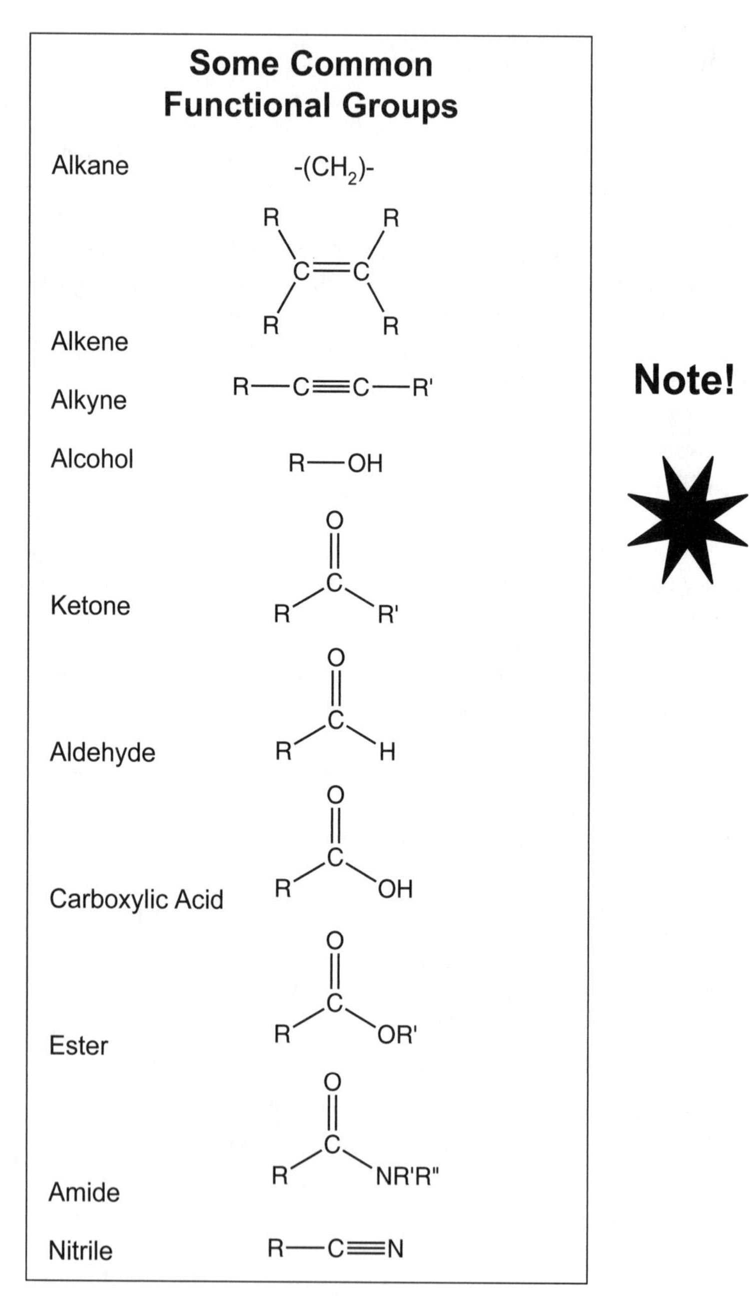
Some Common
Functional Groups
Alkane
-(CH2)-
Alkene
R R C=C R R
Alkyne
R—C≡C—R'
Alcohol
R—OH
Ketone
O C R R'
Aldehyde
O C R H
Carboxylic Acid
O C R OH
Ester
O C R OR'
Amide
O C R NR'R"
Nitrile
R—C≡N
Note!

Atomic Orbitals

An **atomic orbital** (AO) is a region of space about the nucleus in which there is a high probability of finding an electron. For organic molecules, the atomic orbitals of most interest are the s orbital and the p orbitals.

The s orbital is a sphere around the nucleus, as shown below. A p orbital has 2 lobes touching on opposite sides of the nucleus. The three p orbitals are labeled p_x, p_y, and p_z because they are oriented along the x-, y-, and z-axes, respectively . In a p orbital there is no chance of finding an electron at the nucleus—the nucleus is called a **node**.

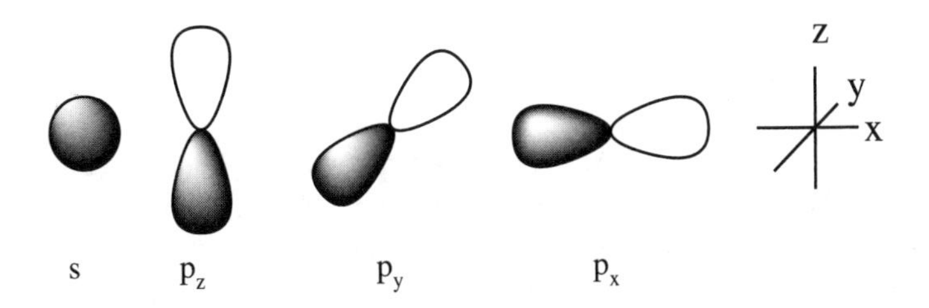

Three principles are used to distribute electrons in orbitals.

1. **"Aufbau" or building-up principle.** Orbitals are filled in order of increasing energy: *1s, 2s, 2p, 3s, 3p, 4s, 3d, 4p, 5s, 4d, 5p, 6s, 4f, 5d, 6p,* etc.
2. **Pauli exclusion principle.** No more than two electrons can occupy an orbital and then only if they have opposite spins.
3. **Hund's rule.** When filling orbitals of equal energy, place one electron in each orbital (using parallel spins) before pairing electrons. (Substances with unpaired electrons are **paramagnetic**—they are attracted to a magnetic field.)

Hybridization and Bonding

A carbon atom must provide four equal-energy orbitals in order to form four equivalent bonds, as in methane, CH_4. It is assumed that the four equivalent orbitals are formed by blending the 2s and the three 2p AO's to give four new **hybrid orbitals,** called sp^3 orbitals. The larger lobe, the "head," having most of the electron density, overlaps with an orbital of its bonding mate to form the bond. The smaller lobe, the "tail," is often omitted when depicting hybrid orbitals. However, at times the "tail" plays an important role in an organic reaction.

The s and p orbitals of carbon can hybridize in ways other than sp^3, as shown below. Repulsion between pairs of electrons causes these hybrid orbitals to have the maximum bond angles. The sp^2 and sp hybrid orbitals induce geometries about the C's as shown below.

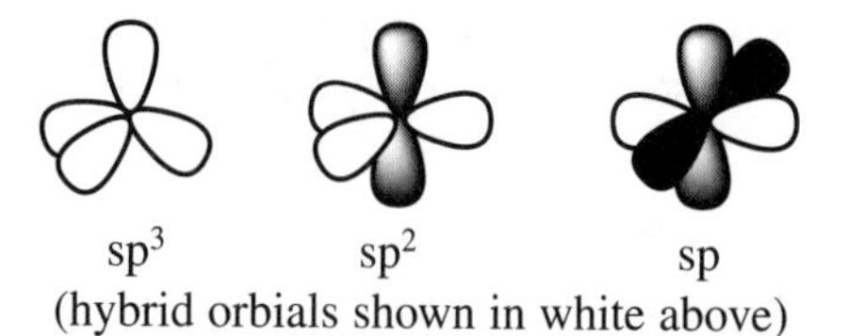

sp^3 sp^2 sp
(hybrid orbials shown in white above)

Head-to-head overlap of AO's gives a **sigma (s) bond**. The **bond angles** (angles between s-bonds) at sp^3 carbons are 109.5°, leading to a tetrahedral geometry, The bond angles at sp^2 carbons are 120°, leading to a trigonal planar geometry, and the bond angles at sp carbons are 180°, leading to a linear geometry. The imaginary line joining the nuclei of the bonding atoms is the **bond axis,** whose length is the **bond length**.

Two parallel p orbitals overlap side-by-side to form a pi (π) bond. The bond axis lies in a nodal plane (plane of zero electronic density). Single bonds are σ bonds. A double bond is one σ and one π bond. A triple bond is one σ and two π bonds.

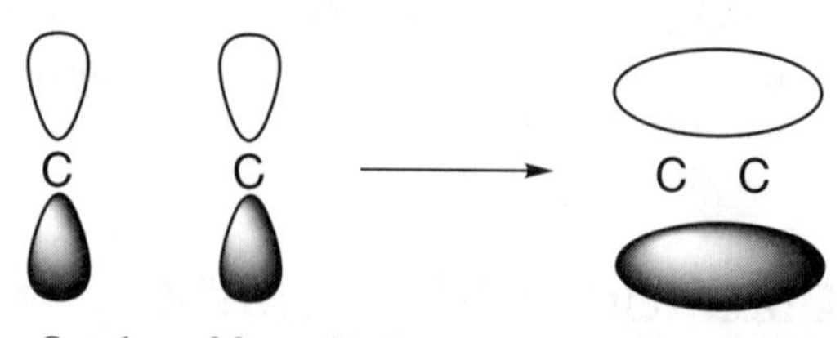

Overlap of 2 p orbitals creates a π bond

Electronegativity and Polarity

The **electronegativity** of an atom is its tendency to attract bonding electrons toward itself. The higher the electronegativity, the more strongly the atom attracts and holds electrons. A **nonpolar** covalent bond exists between atoms having a very small or zero difference in electronegativity. A few relative electronegativities are

F (4.0) > O (3.5) > Cl, N (3.0) > Br (2.8) > S, C, I (2.5) > H (2.1)

A bond formed by atoms of dissimilar electronegativities is called **polar** due to partial charge separation. *The more electronegative element of a covalent bond is relatively negative in charge, while the less electronegative element is relatively positive.* The symbols $\delta+$ and $\delta-$ represent partial charges in polar bonds. These partial charges should not be confused with ionic charges. Polar bonds are indicated by +→; the arrow points toward the more electronegative atom.

The vector sum of all individual bond moments gives the net **dipole moment** of the molecule. H_2O has polar bonds. Since the molecule has a bent shape, the dipoles of the bonds do not cancel and the molecule has a net dipole moment.

Resonance and Delocalized π Electrons

Resonance theory describes species for which a single structure does not adequately describe the species' properties. As an example, consider the cation on the next page (called the **allyl** cation):

A comparison of the calculated and observed bond lengths shows that the 2 C–C bonds are the same length. Neither resonance structure alone can explain this similarity in bond length. When resonance structures form a resonance hybrid, we obtain a structure consistent with the observed bond length. The **resonance hybrid** has some double-bond character between the central carbon and *both* outside carbons. This state of affairs is described by the non-Lewis structure in which dotted lines stand for the partial bonds in which there are delocalized π electrons in an extended π bond created from overlap of p orbitals on each atom. The symbol ↔ denotes resonance, *not equilibrium.*

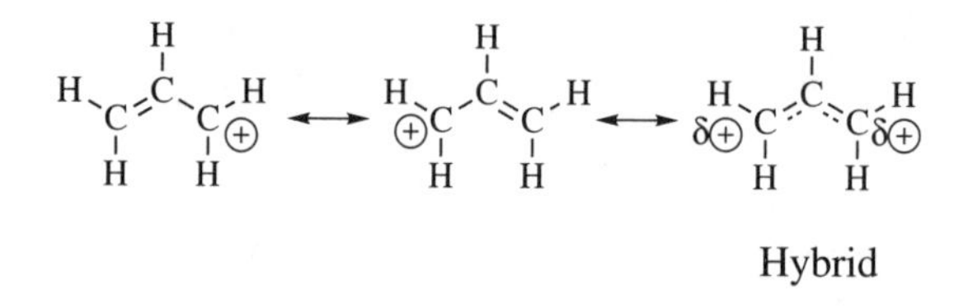

Hybrid

The hybrid is more stable than any single resonance structure. The more nearly equal in energy the contributing structures, the greater the resonance energy. When contributing structures have dissimilar energies, the hybrid looks most like the lowest-energy structure. Contributing structures (a) differ only in positions of electrons (atomic nuclei must have the same positions); (b) must have the same number of paired electrons; and (c) must not place more than 8 electrons on any second period atom. Relative energies of contributing structures are assessed by the following rules.

1. Structures with the greatest number of covalent bonds are most stable. However, for second-period elements (C, O, N) the octet rule must be observed.
2. With few exceptions, structures with the least charge separation are most stable.
3. If all structures have formal charge, the most stable (lowest energy) one has – on the more electronegative atom and + on the more electropositive atom.
4. Structures with like formal charges on adjacent atoms have very high energies.
5. Resonance structures with electron-deficient, positively charged atoms have very high energy, and are usually ignored.

You Need to Know:

- Lewis Dot Structures
- Formal Charges
- Hybridization
- Molecular Geometry
- Resonance

Solved Problems

Problem 1.1 Find the formal charge on each element of $ArBF_3$, and find the net charge on the species.

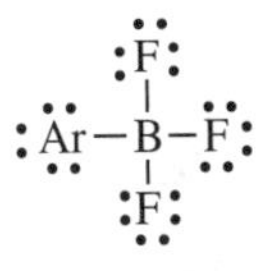

Atom	Group	Unshared Electrons	1/2 Shared Electrons	Formal Charge
F	7	6	1	0
B	3	0	4	-1
Ar	8	6	1	+1
				0 = net charge

Problem 1.2 (a) NO_2^+ is linear, (b) NO_2^- is bent. Explain in terms of the hybrid orbitals used by N.

(a) NO_2^+ (structure). N has two σ bonds, no unshared pairs of electrons and therefore needs two hybrid orbitals. N uses *sp* hybrid orbitals and the σ bonds are linear. The geometry is controlled by the arrangement of the sigma bonds.

(b) NO_2^- (structure). N has two s bonds, one unshared pair of electrons, and therefore, needs three hybrid orbitals. N uses sp^2 hybrid orbitals and the bond angle is about 120°.

Chapter 2
REACTIVITY AND REACTIONS

IN THIS CHAPTER:

- ✔ *Reaction Mechanisms*
- ✔ *Types of Organic Reactions*
- ✔ *Electrophilic and Nucleophilic Reagents*
- ✔ *Thermodynamics*
- ✔ *Bond-Dissociation Energies*
- ✔ *Rates of Reactions*
- ✔ *Chemical Equilibrium*
- ✔ *Transition State Theory and Energy Diagrams*
- ✔ *Brønsted Acids and Bases*
- ✔ *Lewis Acids and Bases*
- ✔ *Solved Problems*

Reaction Mechanisms

The sequence of bond-making and bond-breaking processes in a reaction is called a **mechanism**. A reaction may occur in one step or, more often, by a sequence of several steps. For example, A+B→X+Y may proceed in two steps:

(1) $A \rightarrow I + X$
(2) $B + I \rightarrow Y$

Substances that are formed in early steps and consumed in later steps (such as I in the reaction above) are called **intermediates**. Sometimes the same reactants can give different products via different mechanisms. Intermediates often arise from one of two types of bond cleavage:

Heterolytic (polar) cleavage. Both electrons go with one group, e.g.,

$A{:}B \rightarrow A^+ + {:}B^-$ or $(A{:}^- + B^+)$

Homolytic (radical) cleavage. Each group takes one electron, e.g.,

$A{:}B \rightarrow A\bullet + B\bullet$

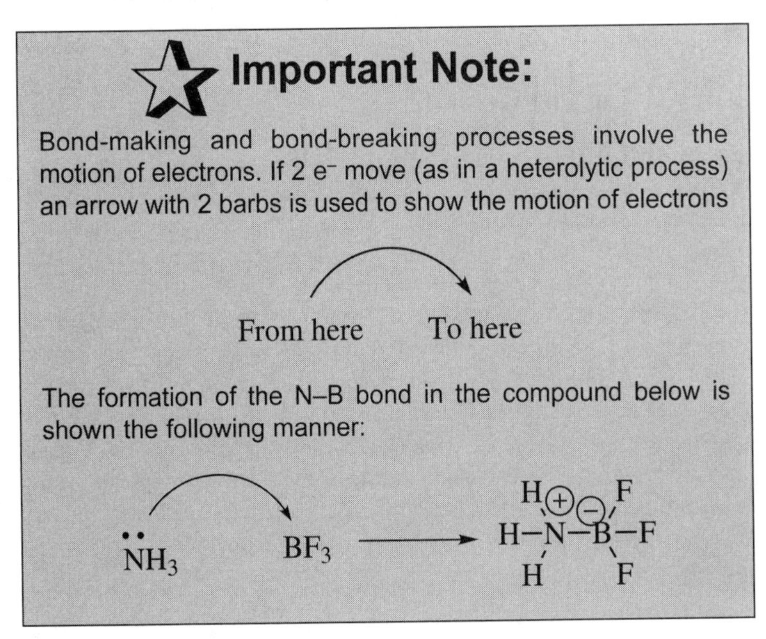

Types of Organic Reactions

Most organic reactions fall into one of the following categories:

1. **Substitution**. An atom or group of atoms in a molecule or ion is replaced by another atom or group.
2. **Addition**. Two molecules combine to yield a single molecule. Addition frequently occurs at a double or triple bond and sometimes at three-membered rings.
3. **Elimination**. This reaction is the reverse of addition. Two atoms or groups are removed from a molecule. If the atoms or groups are taken from adjacent atoms (**β-elimination**), a multiple bond is formed. Removal of two atoms or groups from the same atom (**α-elimination**) produces a **carbene.**
4. **Rearrangement**. Bonds in the molecule are scrambled, converting it to its isomer.
5. **Oxidation-reduction (redox)**. These reactions involve transfer of electrons or change in oxidation number. An increase in the number of H atoms bonded to C and a decrease in the number of bonds to other atoms such as C, O, N, Cl, Br, F, and S signal reduction.

Electrophilic and Nucleophilic Reagents

Reactions generally occur at the reactive sites of molecules and ions. These sites fall mainly into two categories. One category has a high electron density because the site (a) has an unshared pair of electrons or (b) is the δ– end of a polar bond or (c) has C=C π electrons. Such electron-rich sites are **nucleophilic** and the species possessing such sites are called **nucleophiles** or **electron-donors.** The second category (a) is capable of acquiring more electrons or (b) is the δ+ end of a polar bond. These electron-deficient sites are **electrophilic** and the species possessing such sites are called **electrophiles** or **electron-acceptors.** Many reactions occur by covalent bond formation between a nucleophilic and an electrophilic site.

Thermodynamics

The thermodynamics and the rate of a reaction determine whether the reaction proceeds. The thermodynamics of a system is described in terms of several important functions:

(1) ΔH, the change in **enthalpy**, the heat transferred to or from a system. ΔH of a chemical reaction is the difference in the enthalpies of the products and the reactants:

$$\Delta H = [(\text{H of products}) - (\text{H of reactants})]$$

If the bonds in the products are stronger than the bonds in the reactants, energy is released, and ΔH is negative. The reaction is **exothermic.**

(2) ΔS is the change in **entropy**. Entropy is a measure of randomness. The more the randomness, the greater is S; the greater the order, the smaller is S. For a reaction,

$$\Delta S = [(\text{S of products}) - (\text{S of reactants})]$$

(3) ΔG is the change in **free energy.** At constant temperature,

$$\Delta G = \Delta H - T\Delta S \quad (T = \text{absolute temperature})$$

For a reaction to be spontaneous, ΔG must be negative.

Bond-Dissociation Energies

The **bond-dissociation energy** is the energy needed for the endothermic homolysis of a covalent bond A:B $\rightarrow$ A• + •B; ΔH is positive for these reactions. Bond formation, the reverse of this reaction, is exothermic and the ΔH values are negative. Stronger bonds require more energy to break, so they have larger ΔH values. The ΔH of a reaction is the sum of all the (positive) ΔH values for bond cleavages *PLUS* the sum of all the (negative) ΔH values for bond formations.

$$\Delta H = \Delta H_{(\text{bonds broken})} + \Delta H_{(\text{bonds formed})}$$

Rates of Reactions

The rate of a reaction is how quickly reactants disappear or products appear. For the general reaction dA + eB → fC + gD, the rate is given by a rate equation

$$\text{Rate} = k[A]^x[B]^y$$

where k is the **rate constant** at the given temperature, T, and [A] and [B] are molar concentrations (mol/L).

Chemical Equilibrium

Every chemical reaction can proceed in either direction, even if it goes in one direction only to a microscopic extent. A state of **equilibrium** is reached when the concentrations of reactants and products no longer change because the reverse and forward reactions are taking place at the same rate. The **equilibrium constant**, K_{eq}, is defined in terms of molar concentrations as indicated by the square brackets.
For dA+ eB → fC+gD,

$$K_{eq} = \frac{[C]^f[D]^g}{[A]^d[B]^e}$$

The ΔG of a reaction is related to K_{eq} by the expression $\Delta G = -RT \ln K$, where R is the gas constant ($R = 8.314\ \text{Jmol}^{-1}\text{K}^{-1}$) and T is the absolute temperature (in K).

Transition State Theory and Energy Diagrams

When reactants have collided with sufficient energy of activation (E_a or $\Delta G^{\ddagger}$) and with the proper orientation, they pass through a **transition state** in which some bonds are breaking while others are forming. The transition state is the highest energy state between reactants and products. The relationship of the transition state (TS) to the reactants (R) and products (P) is shown by the energy diagram below, which corresponds to a one-step *exothermic* reaction A+B → C+D. At equilibrium, formation of molecules of lower energy is favored. In this reaction, the products (C+ D) are favored. The reaction rate is actually related to the free energy of activation, $\Delta G^{\ddagger}$, where $\Delta G^{\ddagger} = \Delta H^{\ddagger} - T\Delta S^{\ddagger}$.

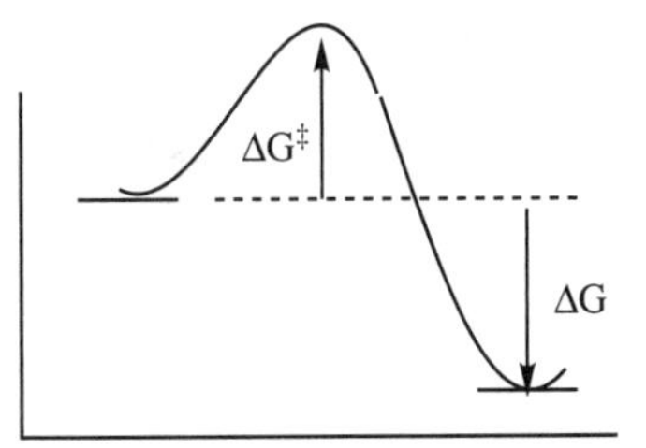

Brønsted Acids and Bases

In the Brønsted definition, an *acid donates a proton* and a *base accepts a proton.* The strengths of acids and bases are measured by the extent to which they lose or gain protons, respectively. In these reactions, acids are converted to their **conjugate bases** and bases to their **conjugate acids.** Acid-base reactions go in the direction of forming the weaker acid and the weaker base. The *strongest* acids have the *weakest* conjugate bases, and the *strongest* bases have the *weakest* conjugate acids. The stronger an acid, the larger its ionization constant K_a and the smaller its pK_a value.

For the acid HA, $HA \rightarrow H^+ + A^-$,

$$K_a = [H^+][A^-]/[HA]$$

$$pK_a = -\log K_a$$

Lewis Acids and Bases

A **Lewis acid (electrophile)** shares an electron pair furnished by a **Lewis base (nucleophile)** to form a covalent (coordinate) bond. The Lewis concept is especially useful in explaining the acidity of an aprotic acid (no available proton), such as BF_3.

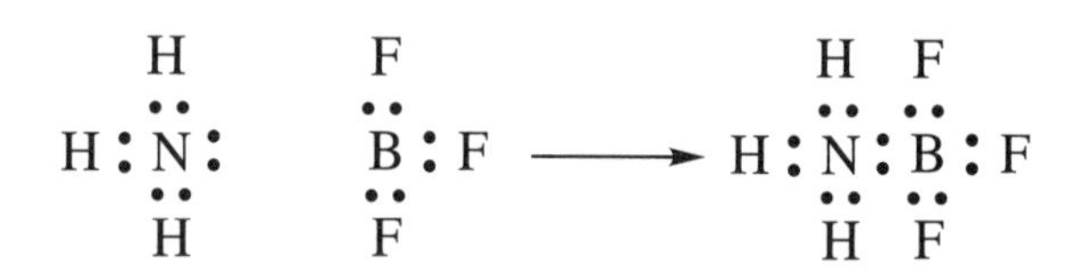

Lewis Base   Lewis Acid

You Need to Know:

- Nucleophiles
- Electrophiles
- ΔG, ΔH, ΔS
- Brønstëd Acids and Bases
- Lewis Acids and Bases

Solved Problems

Problem 2.1 Consider the following sequence of steps:

(1) A ⟶ B

(2) B + C ⟶ D + E

(3) E + A ⟶ 2F

(a) Which species may be described as (i) reactant, (ii) product, and (iii) intermediate?

(b) Write the net chemical equation.

(c) Indicate the molecularity of each step.

(d) If the second step is rate determining, write the rate expression.

(a) (i) A, C; (II) D, F; (iii) B, E.
(b) 2A + C $\longrightarrow$ D + 2F (add steps 1, 2, and 3)
(c) (1) unimolecular, (2) bimolecular, (3) bimolecular.
(d) Rate = k[C][A], since A is needed to make the intermediate, B.

Problem 2.2 Give the conjugate acid of (a) CH_3NH_2, (b) CH_3O^-, (c) CH_3OH, (d) $:H^-$, (e) $:CH_3^-$ (f) $H_2C{=}CH_2$.

(a) $CH_3NH_3^+$, (b) CH_3OH, (c) $CH_3OH_2^+$, (d) H_2, (e) CH_4, (f) $H_3CCH_2^+$.

Chapter 3
Alkanes and Cycloalkanes

In This Chapter:

Saturated Hydrocarbons

Alkanes are hydrocarbons constituting the homologous series with the general formula C_nH_{2n+2}, where n is an integer, for open chain (acyclic) compounds. This formula gives the maximum number of hydrogens per carbon atom, so alkanes are said to be **saturated**. They have only **single bonds**.

A chain of singly bonded C's can be twisted around the C–C bonds into any zigzag shape **(conformation).** Two such arrangements, for four consecutive C's (butane), are shown below.

The two extreme conformations of ethane—called **eclipsed** and **staggered** — can be shown in "wedge-and dash" drawings, "sawhorse" drawings, and in Newman projections. With the Newman projection, we sight along the C—C bond, so that the back C is hidden by the front C. The circle aids in distinguishing the bonds on the front C (touching at the center of the circle) from those on the back C (drawn to the circumference of the circle). In the eclipsed conformation, the bonds on the back C are, for visibility, slightly offset from a truly eclipsed view. The angles between a given C—H bond on the front C and the closest C—H bond on the back C are 60° (in staggered) or 0° (in eclipsed).

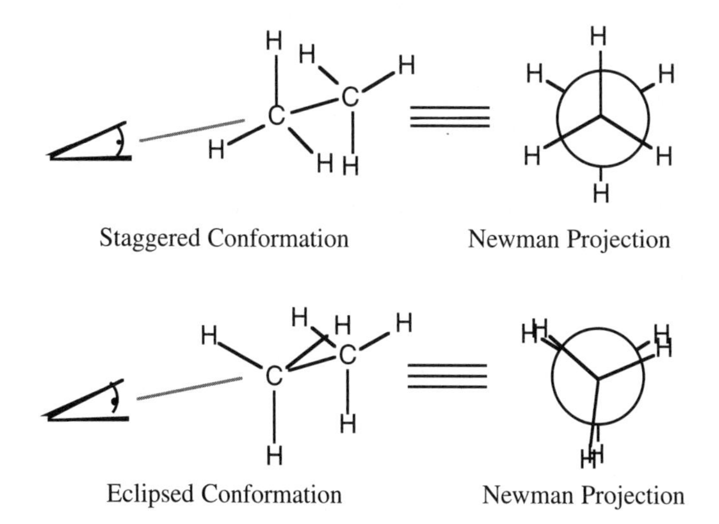

Staggered Conformation

Newman Projection

Eclipsed Conformation

Newman Projection

In the Newman projection of butane, it is possible to see 2 different kinds of staggered conformations. Conformations that place the two CH_3 groups 60° apart are called **gauche** conformations.

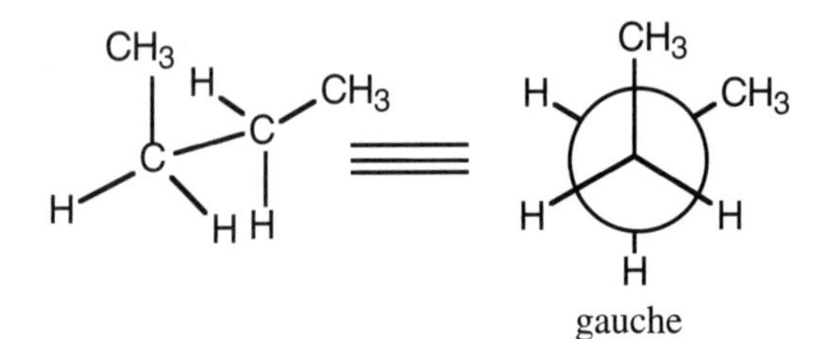

gauche

The gauche conformation is lower in energy than the eclipsed conformation, but higher in energy than staggered conformations that place the CH_3 groups 180° apart, known as the **anti** conformation.

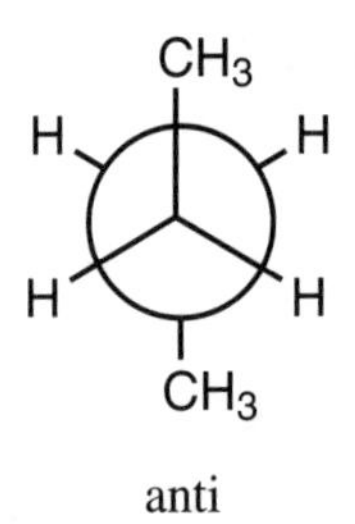

anti

Nomenclature of Alkanes

Alkanes are named by following a set of steps, demonstrated on the following compound:

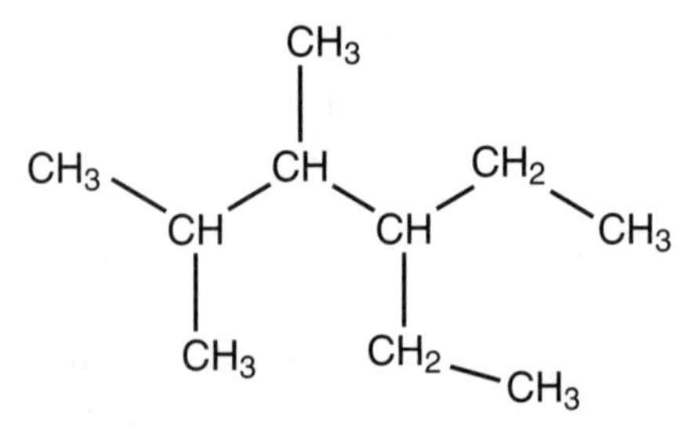

(1) Find the longest unbroken chain ("backbone") of carbon atoms. Find that chain length in the table and add "ane" at the end to denote an alkane.

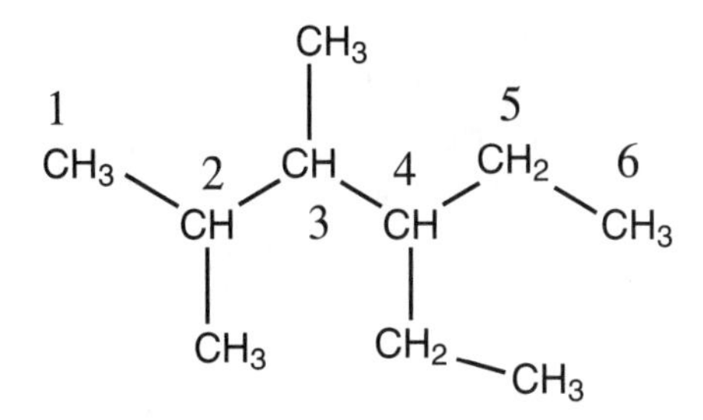

There are 6 carbons in the chain; therefore this is a hexane derivative.

(2) Identify branches. Number the backbone from the end *closest to the first branch.*

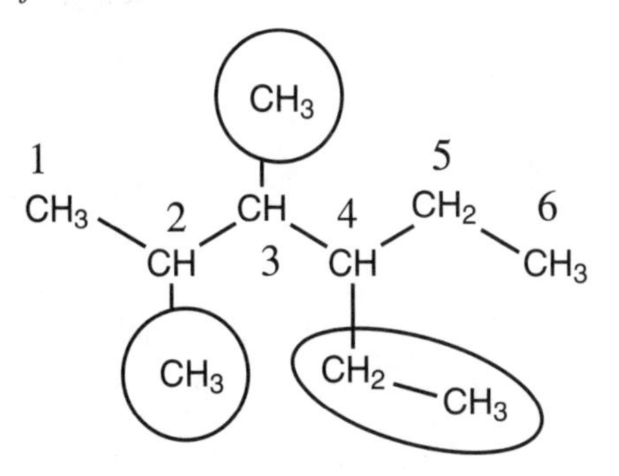

There are CH_3 (methyl) groups at positions 2 and 3, and a CH_3CH_2 (ethyl) group at position 4.

(3) Assemble the name. Put the branches in alphabetical order (ignoring di, tri, etc.). Our example here is 4-ethyl-2,3-dimethylhexane.

The letter *n* (for *normal),* as in n-butane, denotes an unbranched chain of C atoms. The prefixes *sec-* and *tert-* before the name of the group indicate that the H was removed from a secondary or tertiary C, respectively. A **secondary** C has bonds to two other C's, a **tertiary** to three other C's, and a **primary** to two H's and one C. The H's attached to these types of carbon atoms are also called *primary, secondary* and *tertiary* (1°, 2° and 3°), respectively. A **quaternary** C is bonded to four other C's.

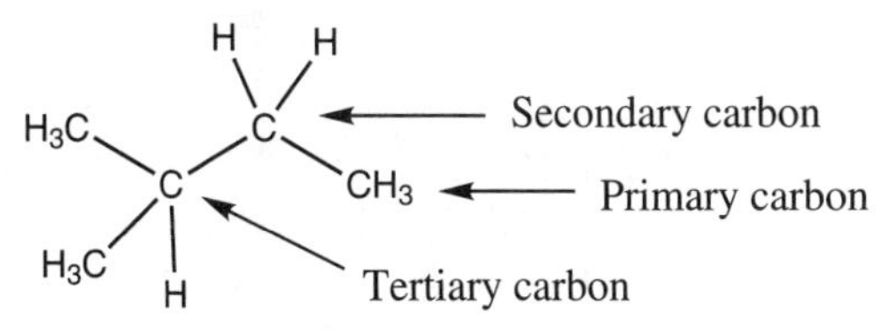

Cyclic hydrocarbons are called **cycloalkanes**. Cycloalkanes have the general formula C_nH_{2n} and are isomeric with alkenes, but, unlike alkenes, they are saturated compounds. They are named by combining the prefix cyclo- with the parent name to indicate the number of carbon atoms in the ring. Two or more substituents are listed alphabetically and are assigned the lowest possible numbers when numbering carbons.

Note!

Common Alkyl Groups:

CH_3-	Methyl
CH_3CH_2-	Ethyl
$CH_3CH_2CH_2$-	Propyl
$(CH_3)_2CH$	Isopropyl
$CH_3CH_2CH_2CH_2$-	Butyl
$(CH_3)_3C$-	tert-Butyl
$CH_3CH_2CH_2CH_2CH_2$-	Pentyl
$CH_3CH_2CH_2CH_2CH_2CH_2$-	Hexyl

Common Hydrocarbons

CH_4	Methane
CH_3CH_3	Ethane
$CH_3CH_2CH_3$	Propane
$CH_3CH_2CH_2CH_3$	Butane
$CH_3CH_2CH_2CH_2CH_3$	Pentane
$CH_3CH_2CH_2CH_2CH_2CH_3$	Hexane

Geometric Isomerism

Atoms in rings cannot rotate completely about their σ bonds, and this leads to *cis-trans* isomers in cycloalkanes. For example, there are 2 isomers of 1,2-dimethylcyclopropane. The isomer in which the 2 CH_3 groups are on the same side of the ring is called the **cis** isomer, and the isomer in which the 2 CH_3 groups are on different sides is called the **trans** isomer.

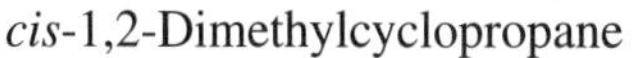
cis-1,2-Dimethylcyclopropane

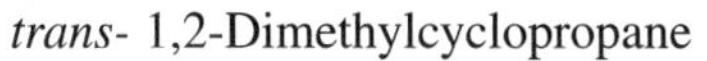
trans- 1,2-Dimethylcyclopropane

There are several ways to draw these structures. In some diagrams, the ring is drawn flat in the plane of the paper, with "wedges" projecting toward the viewer and "dashes" (hashed lines) away from the viewer. Drawings such are the ones below *assume* that there is a carbon atom at each vertex, along with the appropriate number of hydrogens to make each carbon have 4 bonds.

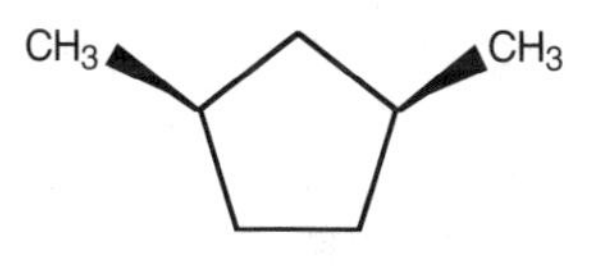

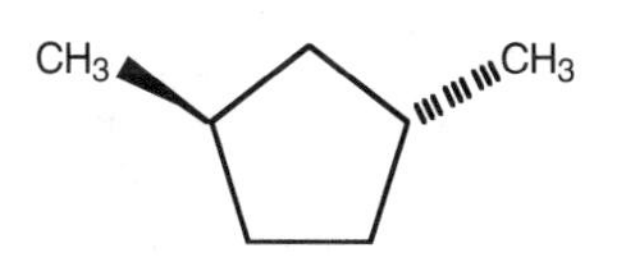

cis-1,3-Dimethylcyclopentane

trans- 1,2-Dimethylcyclopentane

Conformations of Cycloalkanes

There are two common conformations of cyclohexane, the **chair** form and the **boat** form. The chair conformation is more stable than the boat form, since there is less of a tendency for two substituents (in this case hydrogens) to "bump" into each other. Two groups trying to occupy the same space results in unfavorable **steric strain**. Chair conformations and boat conformations interchange through simple bond rotation.

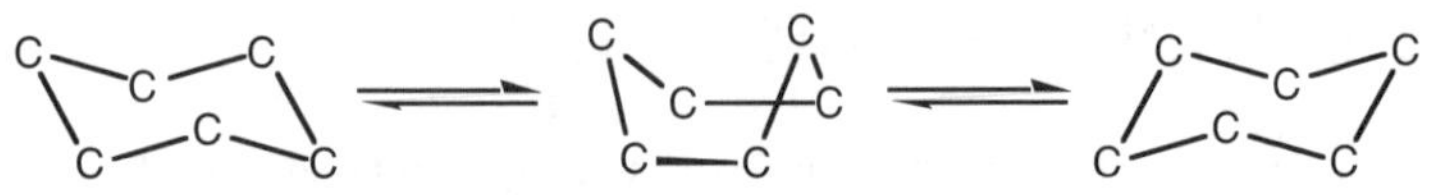

Chair cyclohexane Boat cyclohexane Chair cyclohexane
(hydrogens omitted in each drawing)

You Should Be Able To

- Name hydrocarbons
- Determine which chair conformation is lower energy

Axial and Equatorial Bonds in Chair Cyclohexane

Six of the twelve H's of cyclohexane are **equatorial** (H_e); they project outwards from the ring, forming a belt around the ring perimeter. The other 6 H's are **axial** (H_a); they are perpendicular to the plane of the ring and parallel to each other. Three of these axial H's point up and the other 3 point down. Converting one chair conformer to the other changes *all* of the equatorial bonds in the first conformer, shown as heavy lines, to axial bonds in the second conformer. All of the axial bonds similarly become equatorial bonds. Note how the circled atom below changes from being axial to being equatorial as the first chair converts to a new chair.

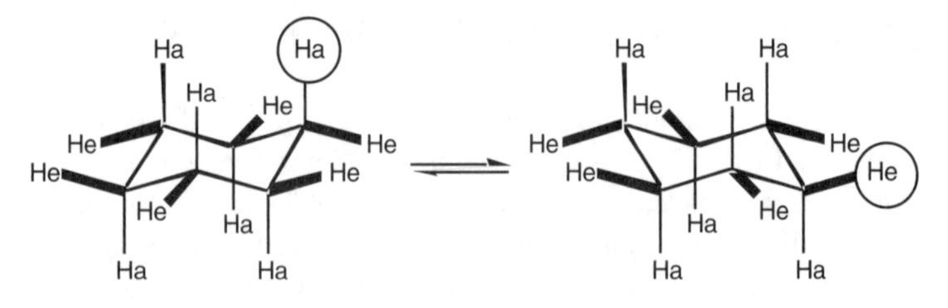

Substituted Cyclohexanes

There are two chair conformations for substituted cyclohexanes such as methylcyclohexane, one in which the CH_3 is axial and one in which the CH_3 is equatorial. The conformer with the axial CH_3 is less stable due to **1,3-diaxial interactions**. The axial CH_3 is closer to the two axial H's (located on the 3rd carbon over) than is the equatorial CH_3 to the adjacent equatorial H's, even those on the adjacent carbons. In general, *substituents prefer the less crowded equatorial position to the more crowded axial position.*

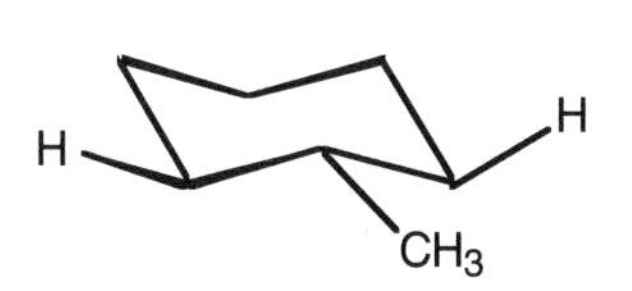

Equatorial CH_3

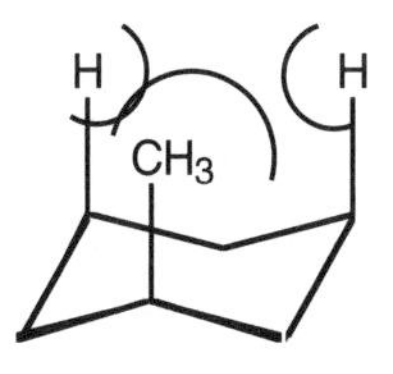

Axial CH_3 (less stable)

Chemical Properties of Alkanes and Cycloalkanes

Alkanes are unreactive except under vigorous conditions.

1. Pyrolytic Cracking [heat (Δ) in absence of O_2; used in making gasoline]:

Alkane → mixture of smaller hydrocarbons

2. Combustion [burning of methane or other alkanes]

$$CH_4 + 2O_2 \rightarrow CO_2 + 2H_2O$$

$$\Delta H \text{ of combustion} = -809.2 \text{ kJ/mol}$$

3. Halogenation

$$RH + X_2 \rightarrow RX + HX$$

(reactivity of X_2: $Cl_2 > Br_2$. I_2 does not react; F_2 destroys the molecule)

Halogenation of alkanes such as methane, CH_4, proceeds by a radical-chain mechanism, as follows:

Initiation Step: Cl–Cl → 2Cl•

The energy required to initiate the reaction comes from ultraviolet (UV) light or heat.

Propagation Steps:

(i) $H_3C–H + Cl• \rightarrow H_3C• + H:Cl$ ΔH= –4 kJ/mol

(ii) $H_3C• + Cl:Cl \rightarrow H_3C–Cl + Cl•$ ΔH= –96 kJ/mol

The sum of the two propagation steps is the overall reaction,

$$CH_4 + Cl_2 \rightarrow CH_3Cl + HCl \qquad = -100 \text{ kJ/mol}$$

In propagation steps, free-radical intermediates, here Cl• and $H_3C•$, are being formed and consumed. Chains terminate when two free-radical intermediates form a covalent bond:

$$Cl• + Cl• \rightarrow Cl_2,\ H_3C• + Cl• \rightarrow H_3CCl,$$
$$H_3C• + •CH_3 \rightarrow H_3C–CH_3$$

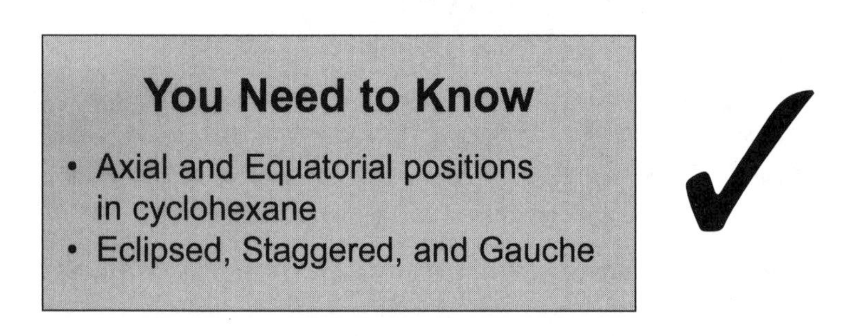

In more complex alkanes, the abstraction of each different kind of H atom gives a different isomeric product. Three factors determine the relative yields of the isomeric product.

(1) **Probability factor.** This factor is based on the number of each kind of H atom in the molecule. For example, in $CH_3CH_2CH_2CH_3$ there are *six* equivalent 1° H's and *four* equivalent 2° H's. The statistical probability of reaction is 6 to 4 (primary to secondary).

(2) **Reactivity of H.** The order of reactivity for different hydrogens in the same molecule is 3° > 2° > 1°.

(3) **Reactivity of X**. The more reactive Cl is less selective and more influenced by the probability factor. The less reactive Br is more selective and less influenced by the probability factor. As summarized by the **reactivity-selectivity principle:** If the attacking species is more reactive, it will be less selective, and the yields will be closer to those expected from the probability factor. An example of this is shown below, in the distribution of 1° and 3° products of halogenation of 2-methylbutane. Bromination, involving the less reactive Br• radical, is much more selective than chlorination.

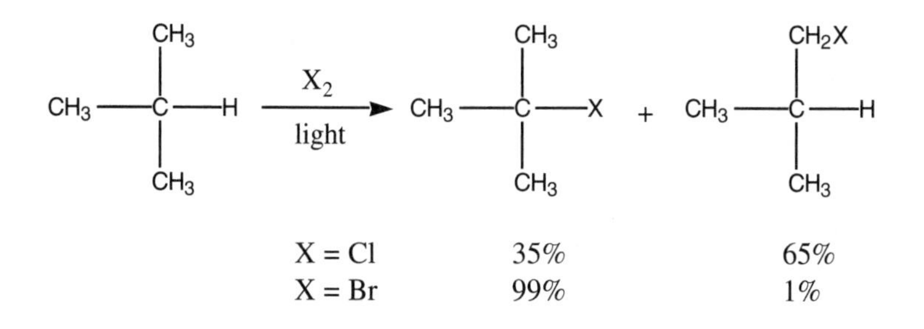

Solved Problem

Problem 3.1 Write structural formulas for the five isomeric hexanes and name them by the IUPAC system.

The isomer with the longest chain is hexane, $CH_3CH_2CH_2CH_2CH_2CH_3$. If we use a five-carbon chain, a CH_3 may be placed on either C2 or C4 to produce 2-methylpentane, or on C3 to give another isomer, 3-methylpentane.

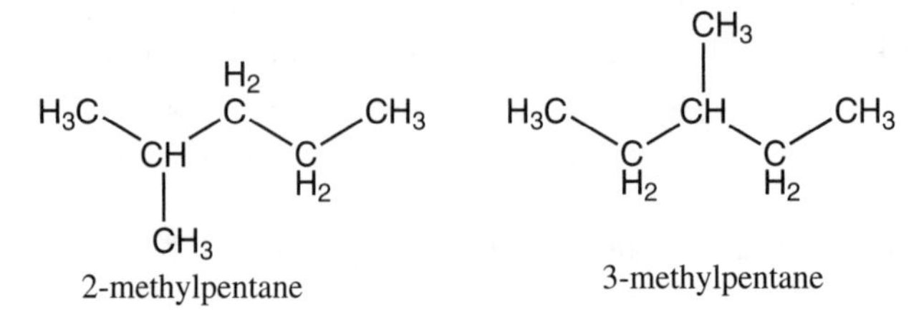

2-methylpentane 3-methylpentane

With a four-carbon chain either a CH_3CH_2 or two CH3's must be added as branches to give a total of 6 C's. Placing CH_3CH_2 anywhere on the chain is ruled out because it lengthens the chain. Two CH_3's are added, but only to central C's to avoid extending the chain. If both CH_3's are introduced on the same C, the isomer is 2,2-dimethylbutane. Placing one CH_3 on each of the central C's give the remaining isomer, 2,3-dimethylbutane.

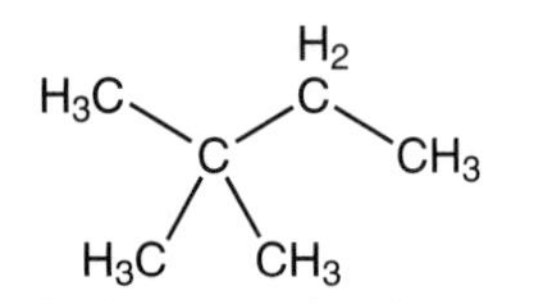

2,2-dimethylbutane

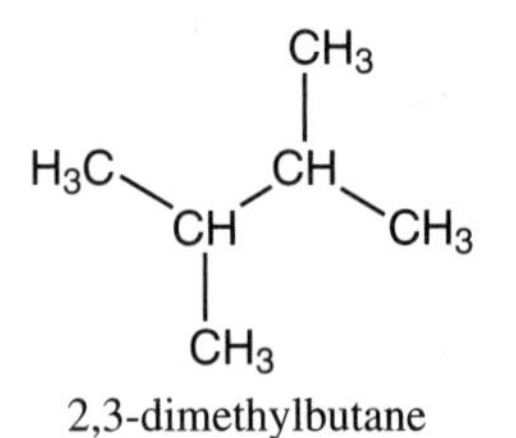

2,3-dimethylbutane

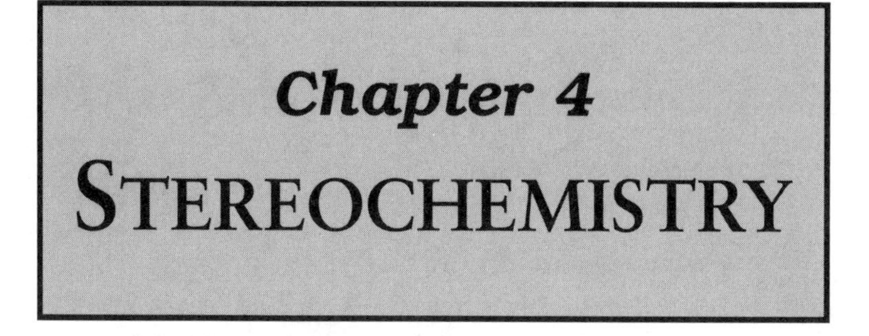

IN THIS CHAPTER:

- ✔ *Stereoisomerism*
- ✔ *Optical Isomerism*
- ✔ *Relative and Absolute Configuration*
- ✔ *Molecules with Multiple Stereogenic Centers*
- ✔ *Solved Problem*

Stereoisomerism

Some structures, be they chemical structures or everyday objects, are not superimposable upon their mirror images. Common examples are your hands and threaded bolts. Other objects, such as marbles or plain white coffee cups, can be perfectly superimposed upon their mirror image. These objects have a *plane of symmetry*: one half of the object is the mirror image of the other.

An object (molecule) that is *not* superimposable on its mirror image is said to be **chiral**. It does *not* possess a plane of symmetry. Structures (molecules) with a plane of symmetry are superimposable upon their mirror images; they are **achiral.**

Most chiral organic molecules contain one or more **stereogenic centers**. These are carbon atoms that are bonded to 4 *different* groups. Examples of chiral compounds, each with a stereogenic center, are shown on the next page:

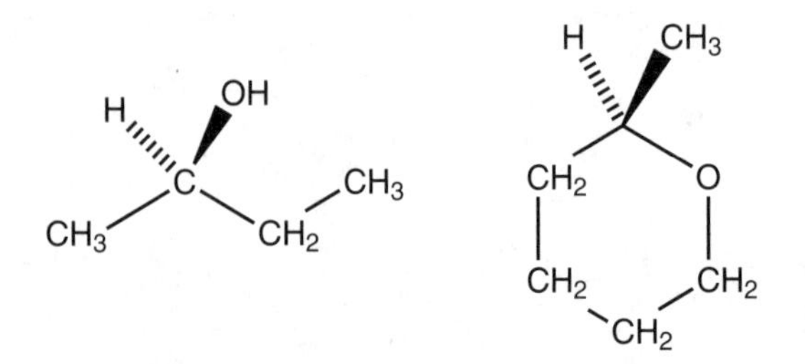

The nonsuperimposable mirror images are called **enantiomers**. A mixture containing amounts of each enantiomer is a **racemic** mixture (a racemate). **Resolution** is the separation of a racemic mixture into individual enantiomers. Enantiomers are a form of isomers called **stereoisomers**. Stereoisomers have the same bonding order of atoms but differ in the way these atoms are arranged in space. Stereoisomers that are not mirror images are called **diastereomers**.

Optical Isomerism

Plane-polarized light (light vibrating in only one plane) passed through a chiral substance emerges vibrating in a different plane. The enantiomer that rotates the plane of polarized light clockwise (to the right) as seen by an observer is **dextrorotatory**; the enantiomer rotating to the left is **levorotatory**. The symbols (+) and (–) designate rotation to the right and left, respectively. Because of this optical activity, enantiomers are called **optical isomers**. A racemic mixture (±) is optically inactive since it does not produce a net rotation of polarized light; the effects of the two enantiomers cancel each other. The **specific rotation** $[\alpha]$ is an inherent physical property of an enantiomer which depends on the solvent used, temperature, and wavelength of light used. It is defined as the observed rotation per unit length of light path per unit concentration (for a solution) or density (for a pure liquid) of the enantiomer; thus,

$$[\alpha] = \frac{\alpha}{l \cdot c}$$

where α is the observed rotation, in degrees
l = length of path, in decimeters (dm)
c = concentration or density, in g/mL.

Relative and Absolute Configuration

Configuration is the 3-dimensional (spatial) arrangement of groups in a stereoisomer. Enantiomers have opposite configurations. For a compound with one stereogenic center to be converted into its enantiomer, bonds must be broken. Configurations may change as a result of chemical reactions. Because stereochemical changes often occur in reactions, it is vital to assign configurations. The sign of rotation cannot be used because *there is no relationship between configuration (spatial arrangement) and sign of rotation.*

The **Cahn-Ingold-Prelog rules** are used to designate the configuration of each chiral C in a molecule in terms of the symbols R and S. These symbols come from the Latin: R from *rectus* (right) and S from *sinister* (left).

Step 1: Groups on the chiral C are assigned *priorities* based on *atomic number* of the atom bonded directly to the C, with higher priority being given to larger atomic numbers.

Step 2: If the first atom is the same in two or more groups, the priority is determined by comparing the next atoms in each of these groups. Thus, ethyl (—CH_2CH_3), with one C and two H's on the first bonded C, has priority over methyl (—CH_3), with three H's on the C.

Step 3: When evaluating the priorities, a double bond counts like <u>two</u> bonds to that element, and a triple bond counts like three bonds to the given atom. For example, for a C=C double bond:

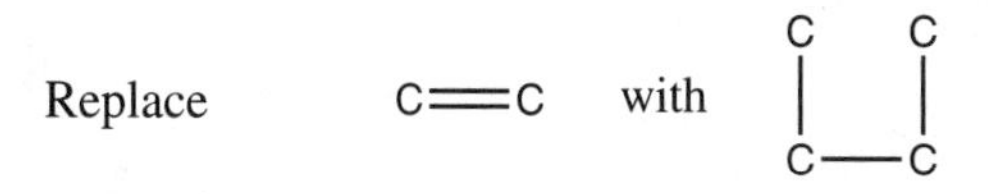

Step 4: Once priorities have been assigned, arrange the molecule so that the lowest-priority group projects behind the plane of the paper, leaving the other three groups projecting forward. Then, for the remaining three groups, if the sequence of decreasing priority, (1) to (2) to (3), is

counterclockwise, the configuration is designated S; if it is *clockwise,* the configuration is designated R. The rule is illustrated for 1-chloro-1-bromoethane below. Both configuration and sign of optical rotation are included in the complete name of a species, e.g., (S)-(+)-1-chloro-1-bromoethane.

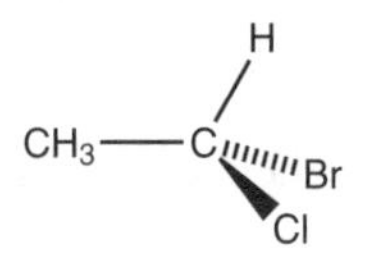

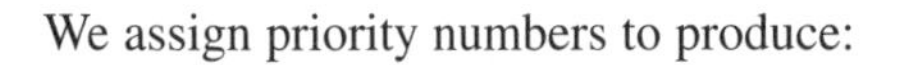
We assign priority numbers to produce:

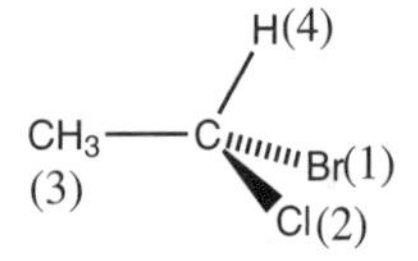

Arranging the molecule to put the lowest priority group behind the plane of the paper:

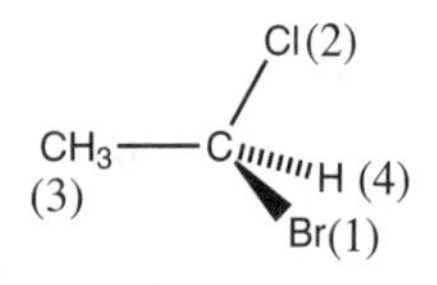

The resulting arrangement shows a counterclockwise sequence, and therefore this molecule has the S absolute configuration.

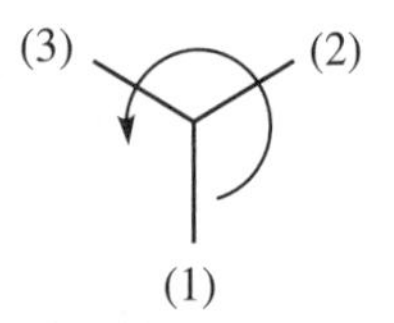

You Need to Know

Definitions:
- Chiral
- Enantiomers
- Racemic
- Absolute Configuration
- Diastereomers
- The R, S Convention

Molecules with Multiple Stereogenic Centers

Molecules containing *n* stereogenic centers can exist as a maximum of 2^n stereoisomers. For example, there are 4 possible stereoisomers of 2,3-dihydroxypentane.

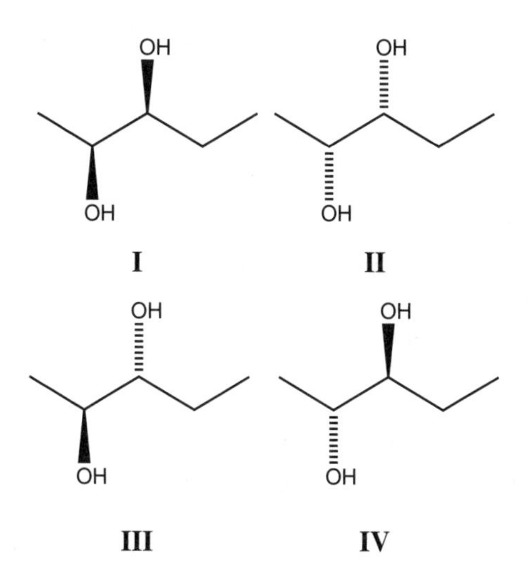

In this example, isomers I and II are enantiomers, and III and IV are enantiomers. The relationship between I and III, as well as between I and IV, is that they are **diastereomers**. Diastereomers are stereoisomers that are not enantiomers.

While it may appear that 2,3-dihydoxybutane can be drawn as 4 different isomers, closer inspection reveals that 2 of these are identical.

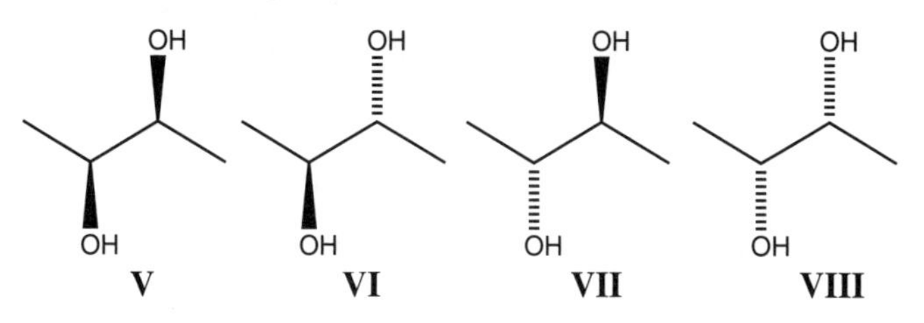

Rotation around the central C–C bond reveals that VI and VII have a plane of symmetry; thus they are therefore achiral and are identical. This type of isomer, an achiral diastereomer, is known as a **meso** compound.

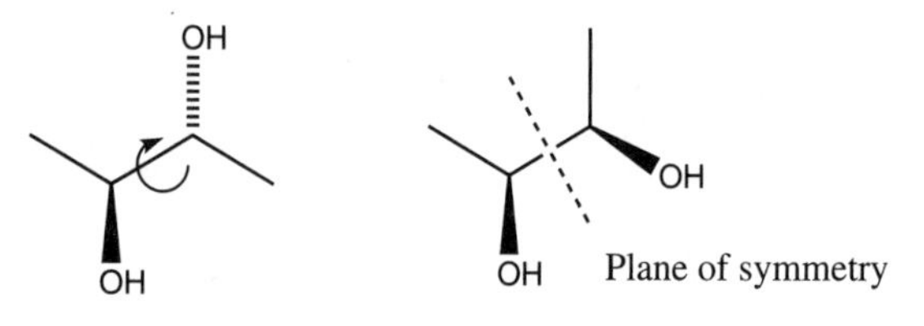

Solved Problem

Problem 4.1 For the following compounds, draw projection formulas for all stereoisomers and point out their R,S specifications, optical activity (where present), and meso compounds:
(a) 1,2,3,4-tetrahydroxybutane
(b) 1-chloro-2,3-dibromobutane

(a) $HOCH_2CHOHCHOHCH_2OH$, with two similar chiral C's, has one *meso* form and two optically active enantiomers.

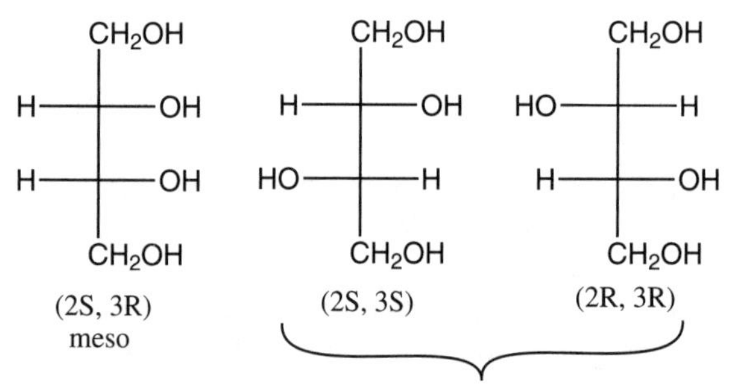

(b) $ClCH_2CHBrCHBrCH_3$ has two different chiral C's. There are four (2^2) optically active enantiomers.

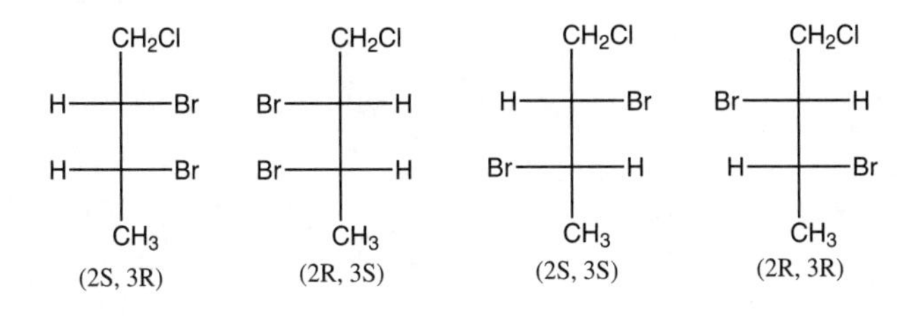

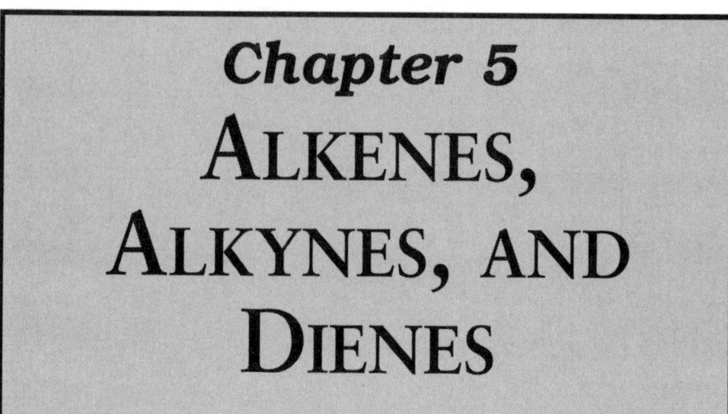

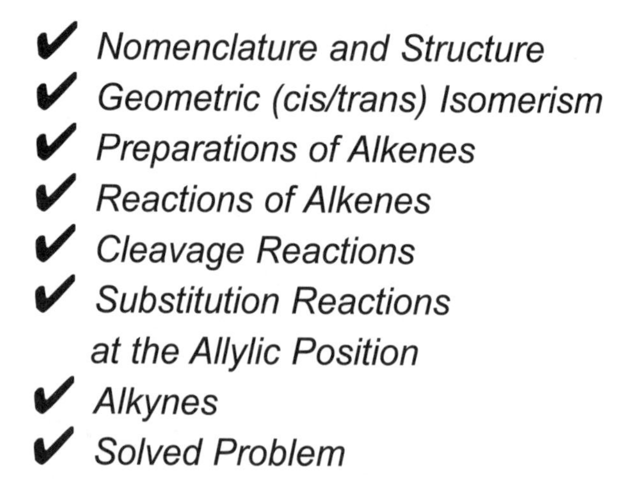

Nomenclature and Structure

Alkenes (olefins) contain a carbon-carbon double bond and have the general formula C_nH_{2n}. These **unsaturated** hydrocarbons are isomeric with the saturated cycloalkanes.

In the IUPAC (International Union of Pure and Applied Chemistry) nomenclature system, the longest continuous chain of C's *containing the double bond* is assigned the parent, with the suffix changed from *-ane* to *-ene*. The chain is numbered so that the position of the double bond is designated by assigning the lower possible number to the first doubly bonded C.

Two important unsaturated groups that have trivial names are CH_2=CH– (Vinyl) and CH_2=CH–CH_2– (Allyl).

Geometric (*cis/trans*) Isomerism

The carbon-carbon double bond consists of a σ bond and a π bond. The π bond has electron density above and below the plane containing the carbon atoms. The π bond is weaker and more reactive than the σ bond because these electrons are more exposed and more weakly bound. Alkenes readily undergo addition reactions. The π bond prevents free rotation about the carbon-carbon double bond and therefore an alkene having two different substituents on each doubly bonded C has geometric isomers. For example, there are two 2-butenes:

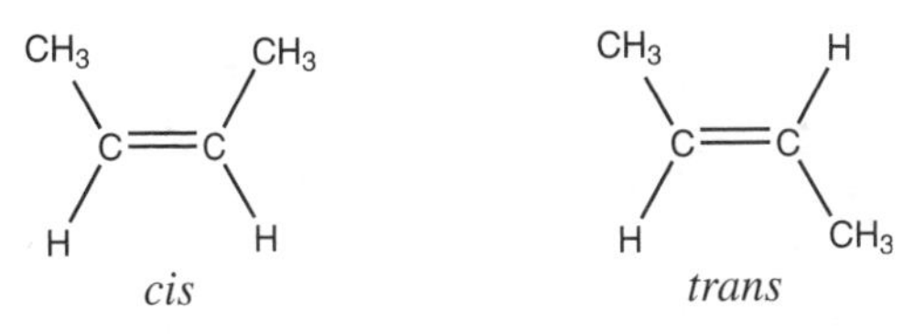

Geometric (*cis/trans*) isomers are stereoisomers because they differ only in the spatial arrangement of the groups. They are diastereomers and have different physical properties (m.p., b.p., etc.). For IUPAC rules, in place of *cis* or *trans*, the letter Z is used if the higher-priority substituents on each carbon are on the same side of the double bond. The letter E is used if they are on opposite sides.

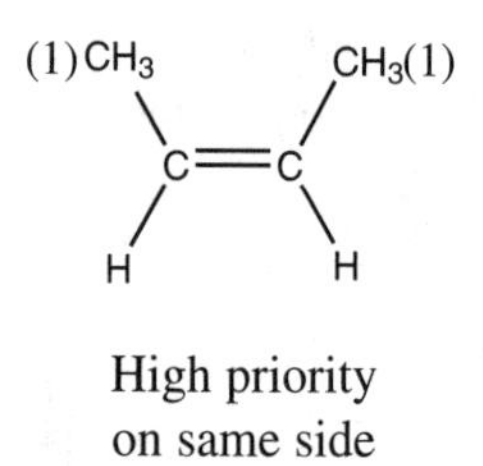

High priority
on same side
= Z

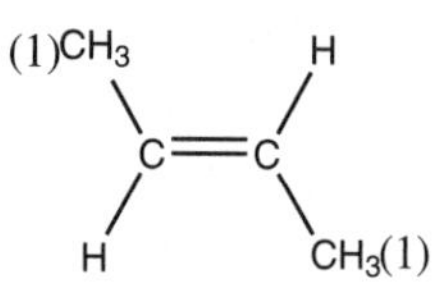

High priority
on opposite sides
= E

Preparations of Alkenes

Elimination. Also called β-eliminations, these reactions constitute the principal laboratory method to make double bonds.

Dehydrohalogenation. Alkenes are typically prepared by treating an alkyl halide with a strong base, often using KOH in ethanol. This reaction proceeds in an antiperiplanar manner (see Chapter 6).

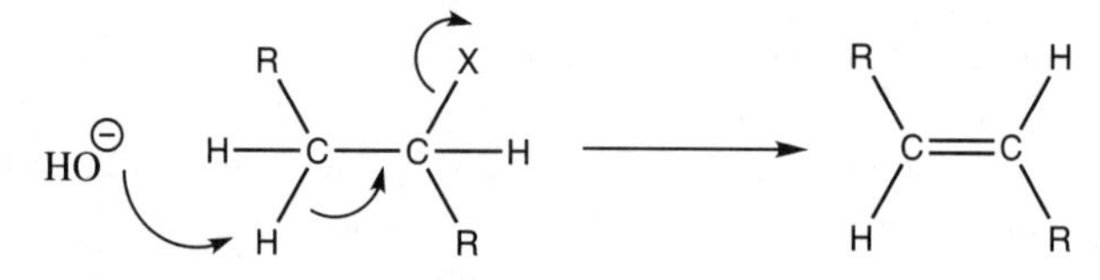

Dehydration. Alcohols are heated with acid to prepare alkenes. Since carbocations can be formed in these reactions, a number of side reactions can interfere.

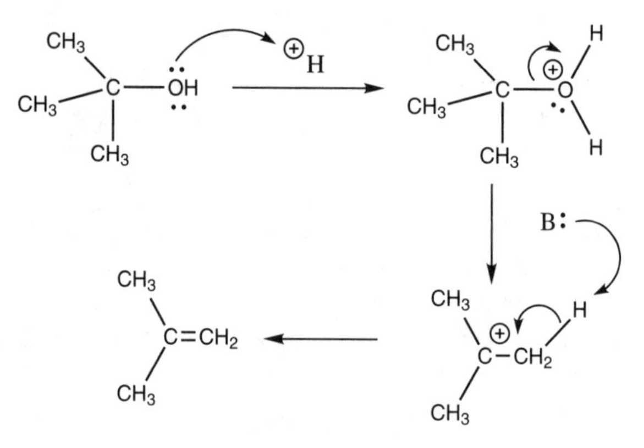

The more groups on the C=C group (i.e., the more substituted the alkene), the more stable is the alkene. The stability of alkenes in decreasing order of substitution by R is:

$$R_2C{=}CR_2 > R_2C{=}CRH > R_2C{=}CH_2\text{ ,}RCH{=}CHR > RCH{=}CH_2$$

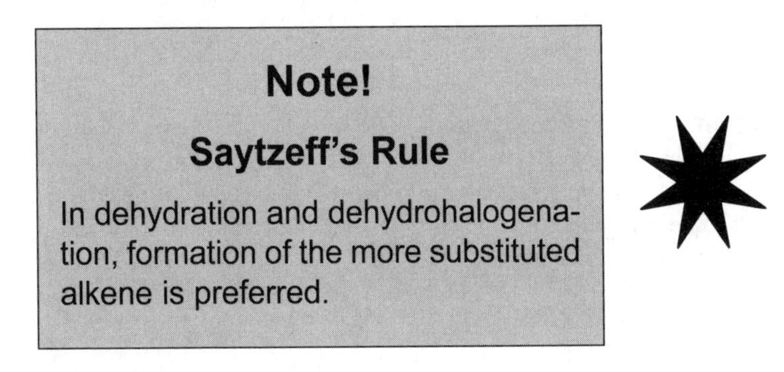

Partial Reduction of Alkynes

Addition of one equivalent of hydrogen to an alkyne produces an alkene. Either E or Z alkenes can be produced, depending on the conditions. Different products are possible because the reactions proceed by different mechanisms.

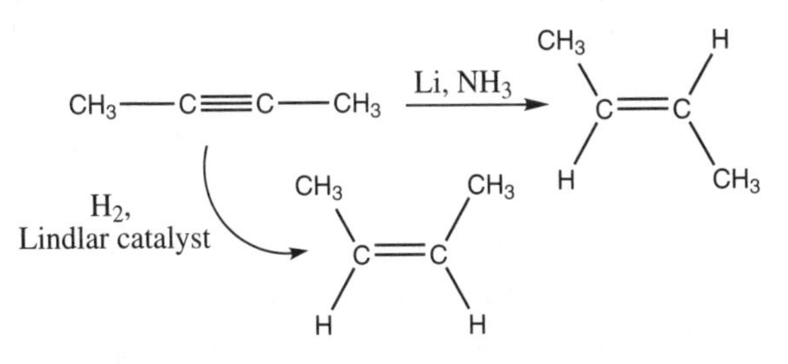

Reactions of Alkenes

Addition of H_2. Alkenes undergo addition reactions at the double bond. The most fundamental of such reactions is the addition of hydrogen. This reaction only occurs in the presence of a catalyst, typically palladium metal supported on activated carbon (Pd/C). Since this reaction involves hydrogen atoms on the surface of the catalyst, both of the hydrogens are added to the same face of the alkene.

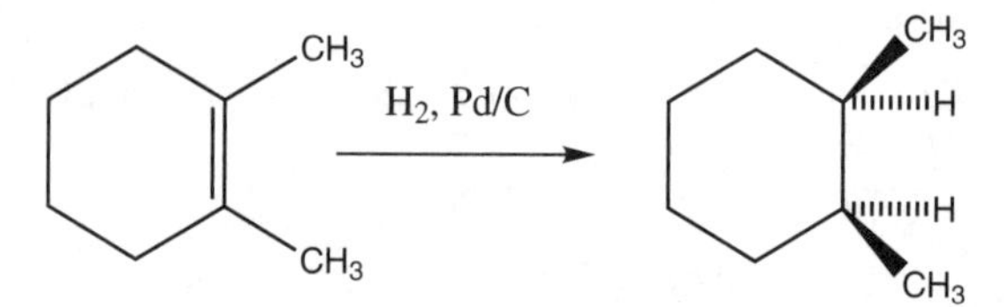

Electrophilic Addition Reactions. The π electrons of alkenes are a nucleophilic site that reacts with electrophiles. The addition of HBr proceeds by the following mechanism. First, the nucleophilic π electrons attack the electrophile, producing a new carbon-hydrogen bond and a carbon cation (carbocation):

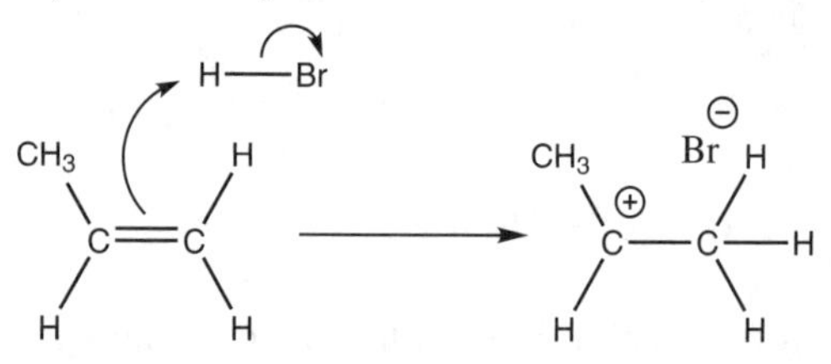

The carbocation (a strong electrophile) is then attacked by bromide ion (a nucleophile), producing the final product.

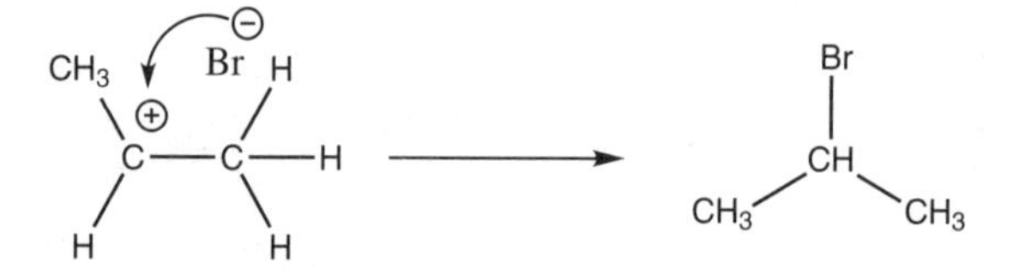

The secondary carbocation shown here is more stable than the primary cation that would result from protonation at the central carbon. Alkyl groups, such as the CH_3 group in the cation above, are electron-donating groups. These electron-donating groups help stabilize the electron-deficient carbocation. As a result, there is a strong preference for the formation of highly substituted cations instead of less-substituted cations.

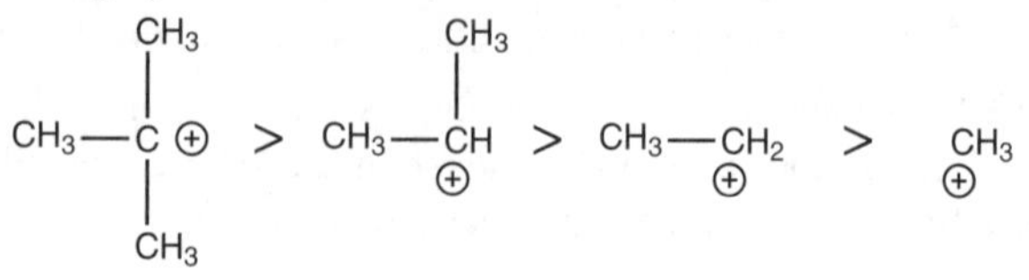

Markovnikov's Rule.

Because of the stability order in carbocations, the addition of HBr or HCl to alkenes results in the addition of the hydrogen to the *less substituted* end of the double bond, and the halogen adds to the *more substituted* end of the double bond.

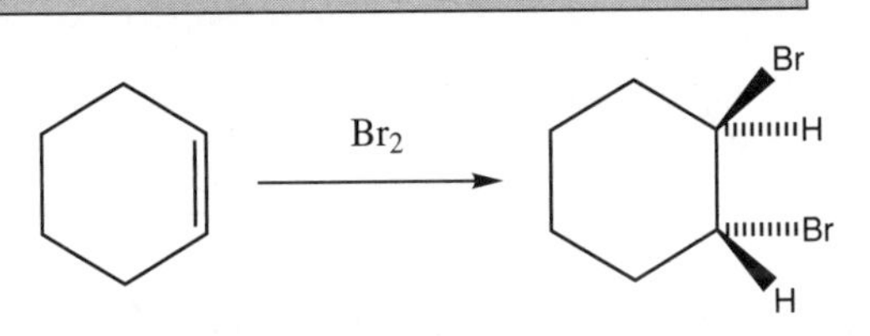

Other addition reactions occur in an analogous fashion. With many electrophiles, there is a very specific stereochemical outcome that is dictated by the reaction mechanism. For example, addition of Br_2 to cyclohexene produces the *trans* isomer of 1,2-dibromocyclohexane:

The reaction proceeds by attack of the nucleophilic π-electrons on Br_2, which produces a bridged bromonium ion.

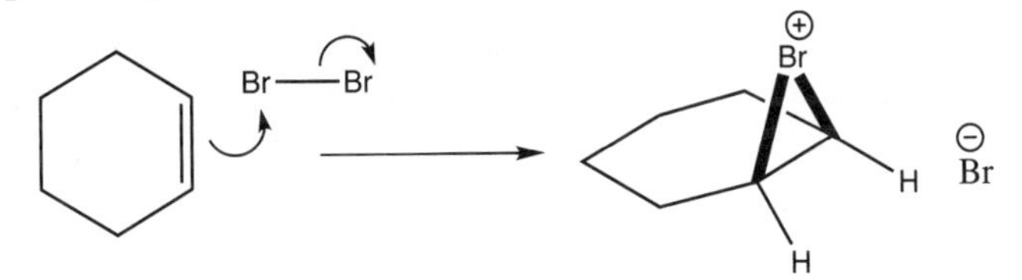

Bromide ion then attacks the bromonium ion from the opposite side of the molecule from the first bromine (in an S_n2 fashion, see Chapter 6) opening the bromonium ion and producing the trans product.

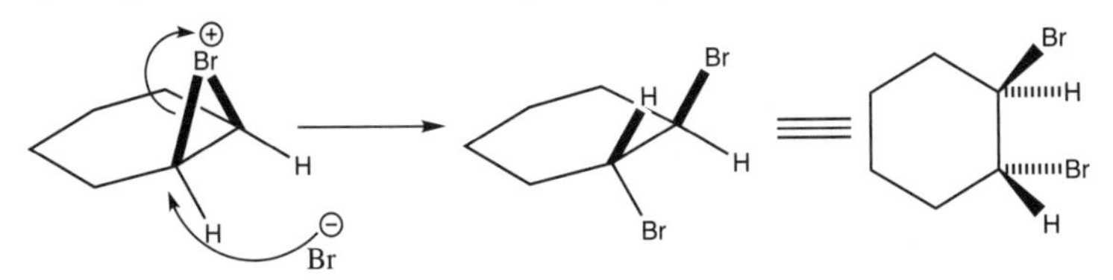

Alkenes also react with *carbenes*, highly reactive 6-electron species. The most common method is to treat the alkene with CH_2I_2 and a zinc-copper alloy. This reaction, known as the **Simmons-Smith** reaction, produces cyclopropanes. The stereochemistry of the alkene is preserved in this reaction. *Trans* alkenes yield *trans* cyclopropanes.

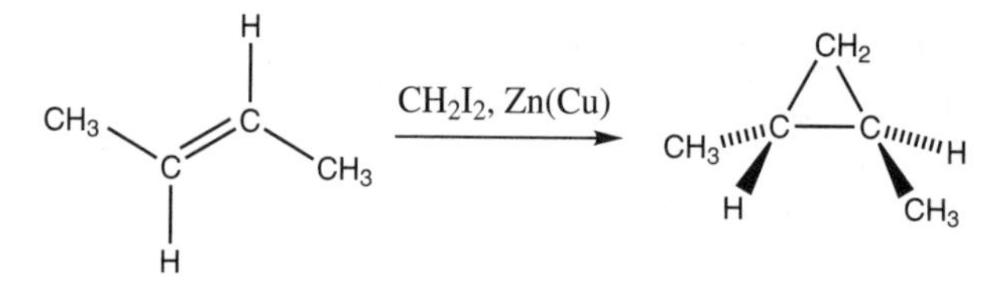

One of the most important reactions of alkenes is hydration. Hydration reactions add water to the double bond, producing alcohols.

Remember!

Treatment of an alkene with water, in the presence of a strong acid, results in hydration, with the orientation of the new H and OH following Markovnikov's rule.

Hydration in an *anti-Markovnikov* manner can be accomplished by a 2-step procedure known as **hydroboration**. In this case, borane (BH_3) adds across the double bond to place the boron on the least substituted carbon. Oxidation with hydrogen peroxide (H_2O_2) results in formation of the alcohol.

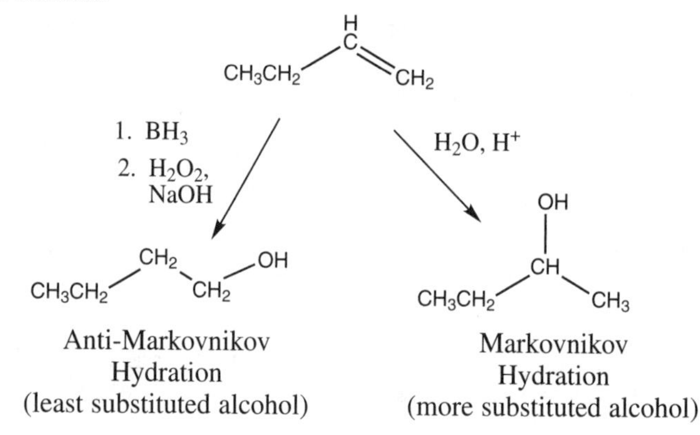

Anti-Markovnikov
Hydration
(least substituted alcohol)

Markovnikov
Hydration
(more substituted alcohol)

Cleavage Reactions

Alkenes can be cleaved by oxidation with any of several different reagents. Oxidation to a 1,2-diol (a dialcohol) with OsO_4 followed by cleavage with periodate is one common 2-step route:

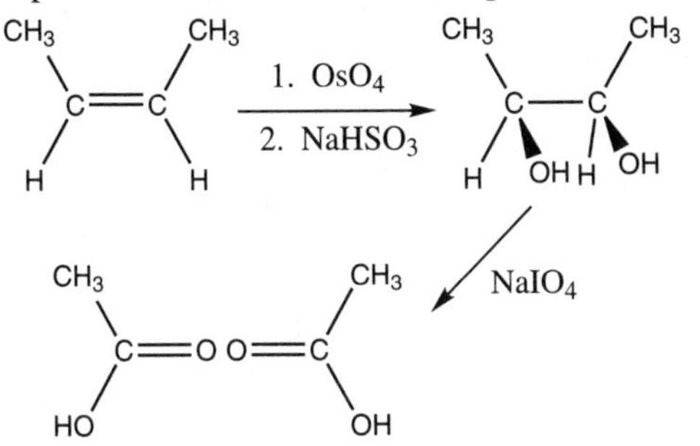

Ozonolysis. Ozone (O_3) reacts with alkenes to form unstable *ozonides*, which must then be reduced (CH_3SCH_3, Ph_3P, or Zn) or oxidized (H_2O_2) to produce a stable product. The products are ketones, aldehydes, or carboxylic acids, depending on the number of alkyl groups on the terminal carbon and whether the ozonide is oxidized or reduced.

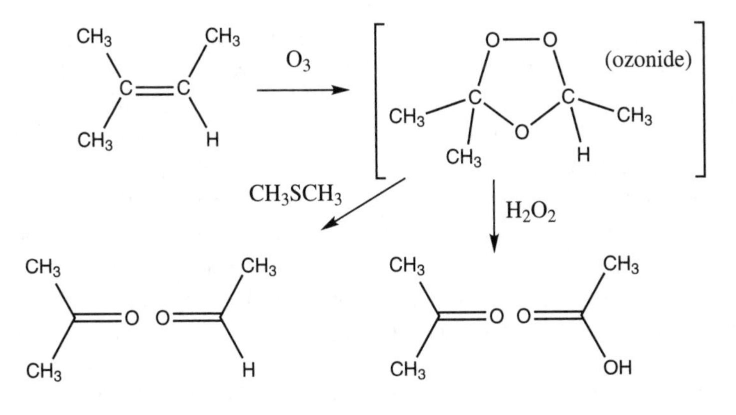

Substitution Reactions at the Allylic Position

sp^3 Hybridized carbon atoms that are directly bonded to one of the sp^2 carbons of a double bond are known as *allylic* carbons. These sites are particularly reactive, since any intermediates formed here are stabilized by **resonance.**

Allylic sites are easily halogenated with N-bromosuccinimide (NBS) in a radical reaction. Peroxides or light (hν) are required to initiate the reaction.

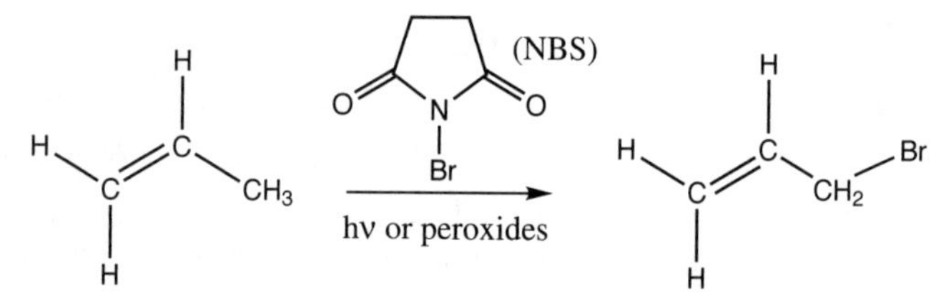

Conjugation and Resonance. Compounds with more than one double bond are **conjugated** if there is an alternating arrangment of double and single bonds, with no sp^3 carbons in between. This arrangement of double bonds is more stable than the unconjugated arrangement. **Resonance** involves different ways to allocate electrons within an extended π system. For example, the compound acrolein (on the next page) has a conjugated arrangement of a C–C double bond and a C–O double bond, giving it 4 electrons in the π system. There are 3 ways these electrons can be allocated among the 4 sp^2-hybridized atoms, according to the three resonance forms shown. Note that all of the nuclei remain in the same place and that the σ bond framework does not change. The most accurate picture of the molecule is a **hybrid** of all of these individual resonance forms. Essentially, the π electrons are delocalized among the 4 sp^2-hybridized atoms. Resonance is an extremely important, powerfully stabilizing influence.

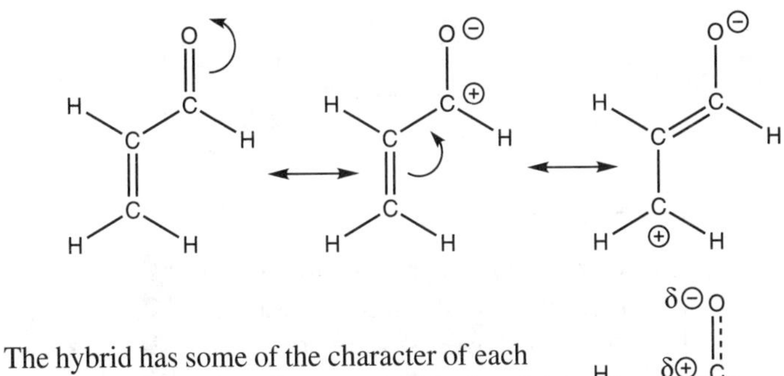

The hybrid has some of the character of each resonance form, with some double bond character between each of the 4 sp^2 atoms, and partial charges on several atoms as well:

Reactions of Conjugated Dienes. Typical of conjugated dienes, 1,3-butadiene undergoes both 1,2- and 1,4-addition, as illustrated with the addition of HBr. The product distribution is strongly dependent on reaction conditions.

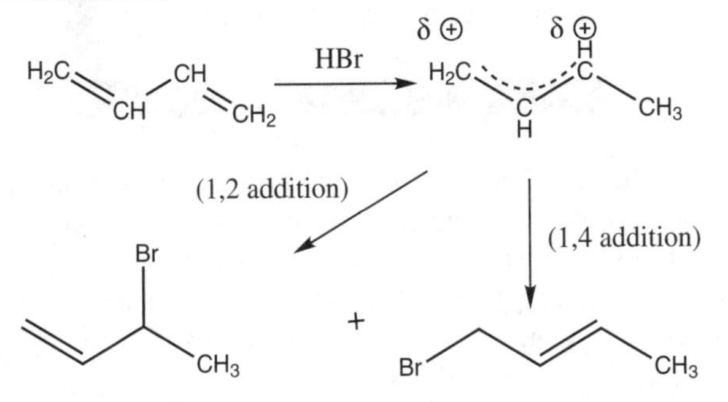

The resonance hybrid for the cation intermediate can be attacked at either end, leading to either one (or both) of the products. At low temperature, 1,2-addition predominates and at higher temperature 1,4-addition does.

Cycloadditions. A useful synthetic reaction is cycloaddition of an alkene, called a **dienophile,** to a conjugated diene by 1,4-addition. This reaction is known as the **Diels-Alder** reaction.

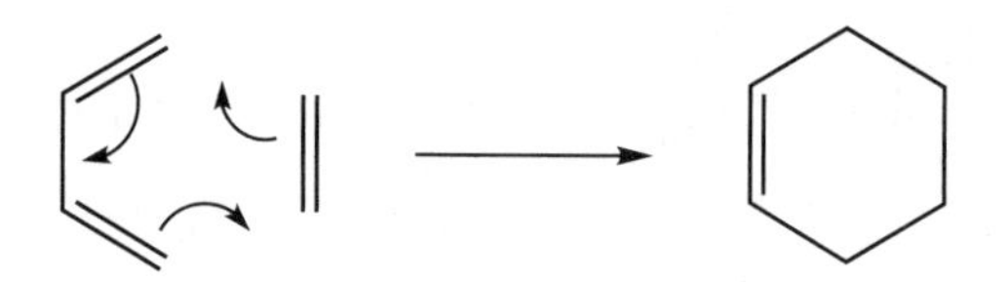

A **kinetically controlled** reaction is one whose major product is formed through the lowest-energy transition state (lowest $\Delta G^{\ddagger}$). A **thermodynamically controlled** reaction is one whose major product has the lower (more negative) ΔH of reaction. Reactions may shift from kinetic to thermodynamic control with increasing temperature when the formation of the rate-controlled product is reversible and chemical equilibrium can be established.

The Diels-Alder reaction is a **concerted** process; no intermediates are formed as the reaction progresses. Due to the concerted nature of these reactions, the stereochemistry of the starting materials is retained in the products. Typical Diels-Alder reactions involve relatively electron-deficient dieneophiles, usually alkenes with carbonyl groups directly bonded to the sp^2 carbons.

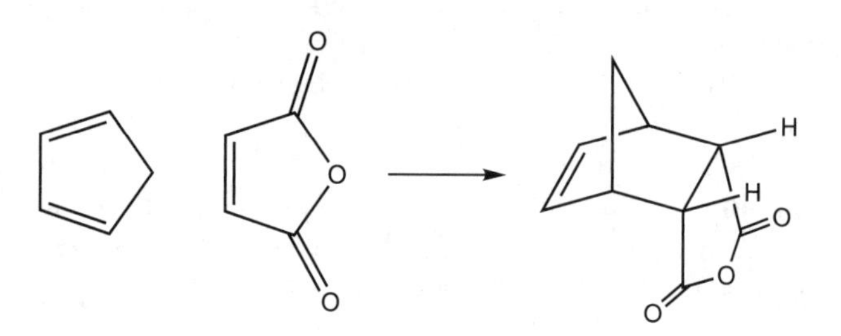

Alkynes

Alkynes (C_nH_{2n-2}) have a carbon-carbon triple bond. These compounds are named by combining the parent name with the suffix -yne. Acetylene (ethyne based on IUPAC nomenclature rules), C_2H_2, is a linear molecule in which each C uses two sp hybrid orbitals to form two σ bonds at a 180° angle. The unhybridized p orbitals form two π bonds. The shorter bond length makes for greater orbital overlap and a stronger bond.

Preparation of Alkynes

Dehydrohalogenation of 1,2 or 1,1 Dihalides. Treatment of dihalides with strong base results in the formation of an alkyne. The vinyl (alkenyl) halide requires the stronger base sodamide ($NaNH_2$).

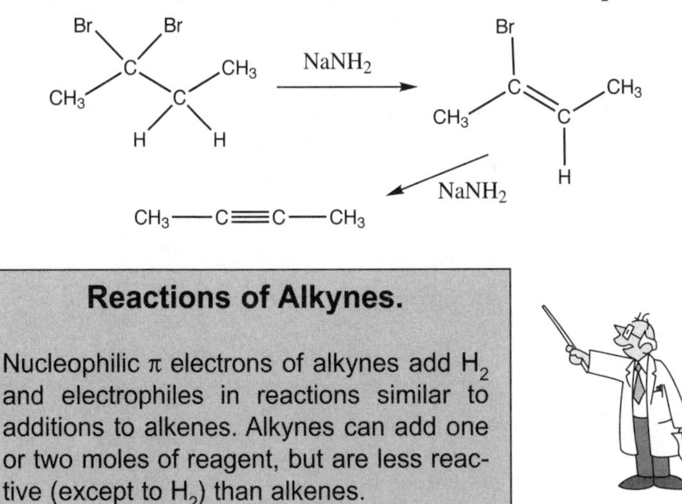

Reactions of Alkynes.

Nucleophilic π electrons of alkynes add H_2 and electrophiles in reactions similar to additions to alkenes. Alkynes can add one or two moles of reagent, but are less reactive (except to H_2) than alkenes.

Hydrogenation. Alkynes react rapidly with 2 equivalents of hydrogen over a Pd/C catalyst to produce alkanes. Partial hydrogenation (to *cis*-alkenes) is possible if a poisoned catalyst (a Lindlar catalyst) is used. Alternatively, reduction with sodium in ammonia produces the *trans* isomer.

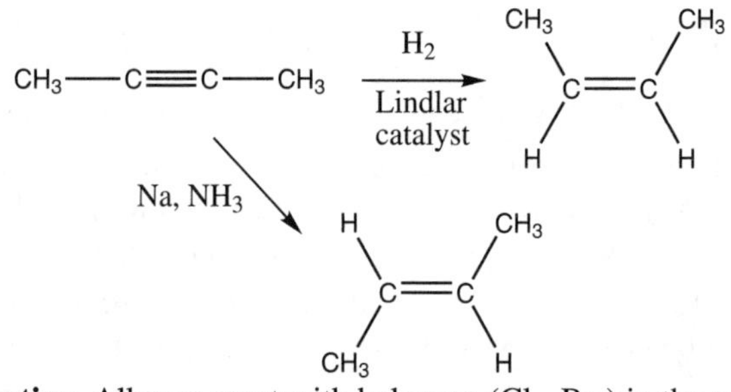

Halogenation. Alkynes react with halogens (Cl_2, Br_2) in the same manner as do alkenes. The first equivalent adds in an anti fashion, producing E dihalo alkenes.

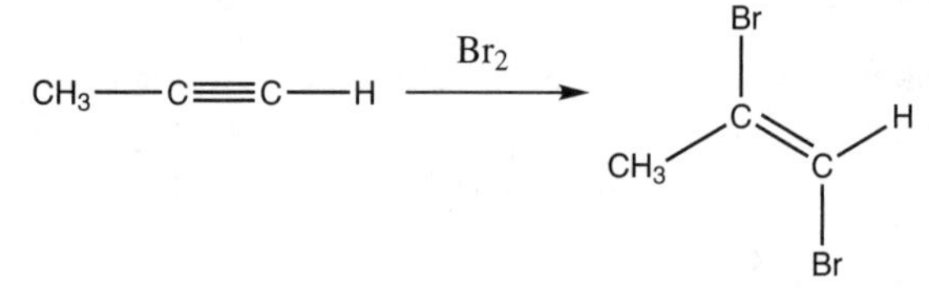

Hydration. Addition of water to an alkyne produces an **enol**. These compounds are unstable relative to their carbonyl-containing isomers. Hydration using $HgSO_4$ produces ketones, and 2 isomers are formed if the alkyne is unsymmetrical. Methyl ketones are favored from the reaction of terminal alkynes, as shown below.

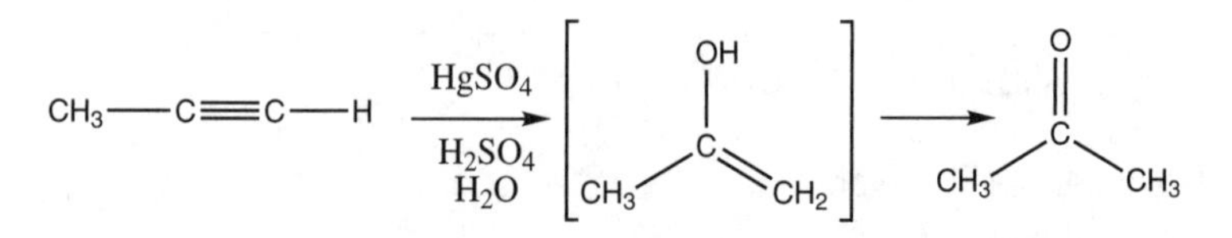

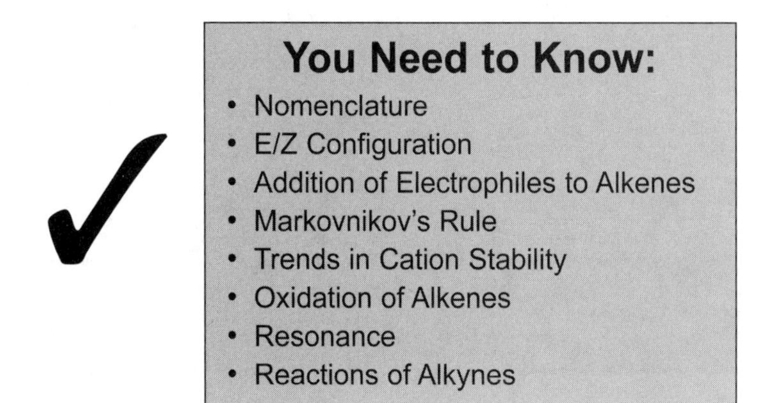

Solved Problem

Problem 5.1 Provide products for the following reaction.

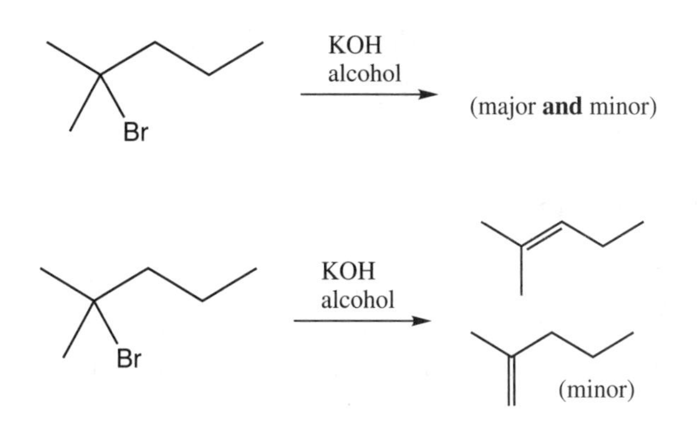

Chapter 6 ALKYL HALIDES

IN THIS CHAPTER:

- ✔ *Nomenclature*
- ✔ *Synthesis of Alkyl Halides*
- ✔ *Chemical Reactions of Alkyl Halides*
- ✔ *Solved Problems*

Nomenclature

Alkyl halides have the general formula RX, where R is an alkyl or substituted alkyl group and X is any halogen atom (F, Cl, Br, or I). These compounds are named as the parent hydrocarbon, using fluoro-, chloro-, bromo-, or iodo- as the substituent names. For example,

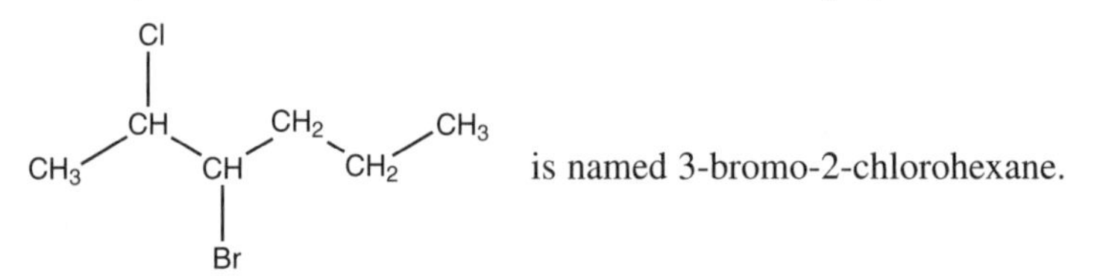

is named 3-bromo-2-chlorohexane.

Classification is based on the structural features: RCH_2Br is 1°, R_2CHBr is 2°, and R_3CBr is 3°.

Synthesis of Alkyl Halides

Alkyl halides are usually prepared from alcohols or from alkenes, although they can be prepared directly from alkanes.

1. Halogenation of alkanes with Cl_2 or Br_2. These reactions are usually initiated by light (hν) and involve radical intermediates. Several steps are involved, and summarized as follows:

Initiation (produces 2 halogen radicals):

$$X_2 \xrightarrow{h\nu} 2\ X\bullet$$

Propagation (produces product and continues the chain):

$$\begin{cases} R\text{-}H + X\bullet \longrightarrow R\bullet + HX \\ R\bullet + X_2 \longrightarrow RX + X\bullet \end{cases}$$

Termination (reactions that destroy radicals):

$$2\ R\bullet \longrightarrow R\text{-}R$$

$$R\bullet + X\bullet \longrightarrow R\text{-}X$$

2. From alcohols (ROH) with HX, PX_3, or $SOCl_2$. These reactions are similar to the nucleophilic substitution reactions discussed later in this chapter.

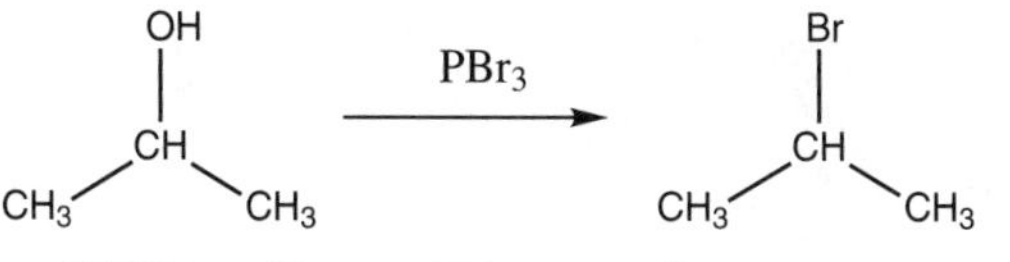

3. Addition of HX to alkenes. These reactions, which tend to follow Markovnikov's rule, are discussed more fully in Chapter 5.

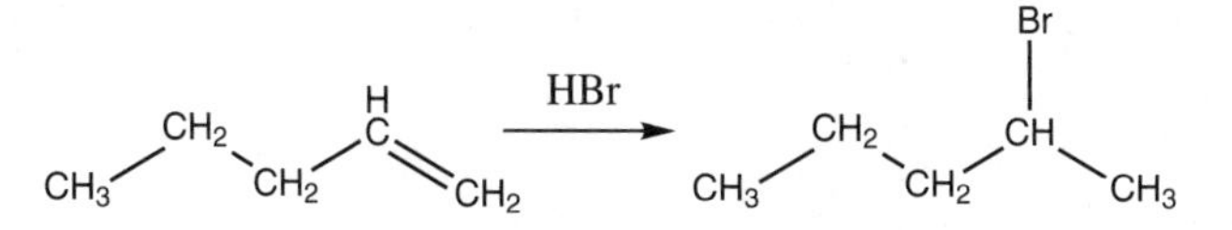

4. By reaction of alkenes with X_2 (X = Br, Cl) to give 1,2-dihalides. These electrophilic addition reactions are discussed more fully in Chapter 5.

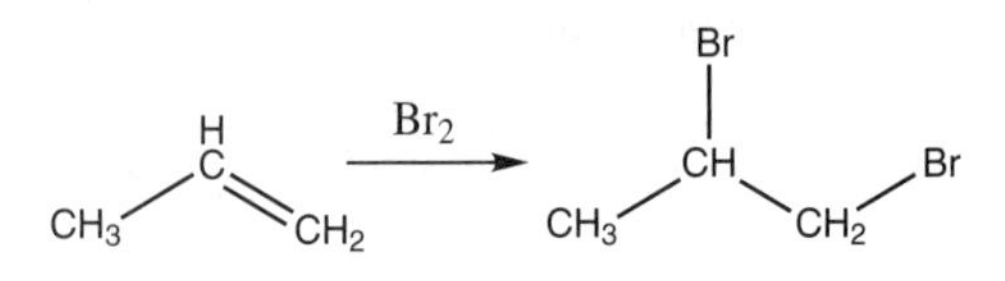

Chemical Reactions of Alkyl Halides

Alkyl halides react with nucleophiles and with strong bases. Reactions with nucleophiles result in **substitution**, while **elimination** reactions result from reactions with bases.

Nucleophilic Substitution. Substitution reactions are reactions in which one group (a halogen, for example) is replaced by another group. Alkyl halides are electrophiles, so they react with nucleophiles to give substitution products. The halide ion that is displaced by the incoming nucleophile is called the **leaving group.**

Nucleophile + Substrate → Product + Leaving Group

Sulfonates are excellent leaving groups—much better than the halides. One of the best leaving groups (better than Br^-) is $CF_3SO_3^-$, called **triflate.**

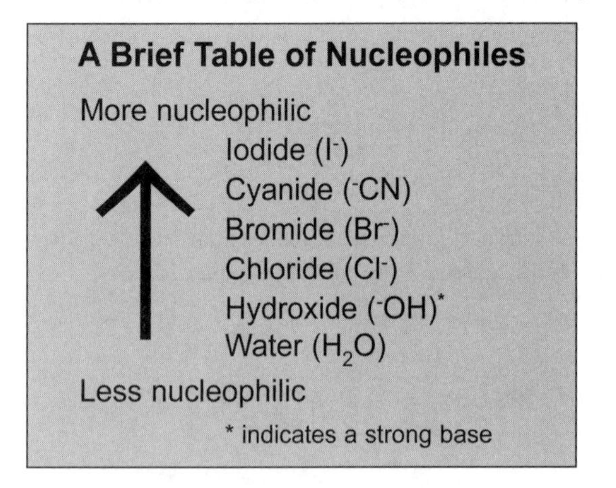

A Brief Table of Nucleophiles

More nucleophilic

Iodide (I^-)
Cyanide (^-CN)
Bromide (Br^-)
Chloride (Cl^-)
Hydroxide (^-OH)*
Water (H_2O)

Less nucleophilic

* indicates a strong base

S_n1 and S_n2 Mechanisms

The two major mechanisms of nucleophilic substitution are S_n1 and S_n2. The "S" means substitution, the "n" refers to the fact that a nucleophile is involved, and the "1" and "2" indicate the order of the reaction. S_n1 is a first-order reaction mechanism, meaning that only a single molecule is involved in the transition state for the rate-determining step. S_n2 is a second-order reaction in which 2 molecules (alkyl halide and nucleophile) come together in the transition state for the slow (rate determining) step.

The mechanism of the S_n1 reaction is shown below. Upon heating the alkyl halide, the C–X bond breaks, creating a carbocation and the halide ion leaves. The carbocation then reacts with a nucleophile to produce the substitution product.

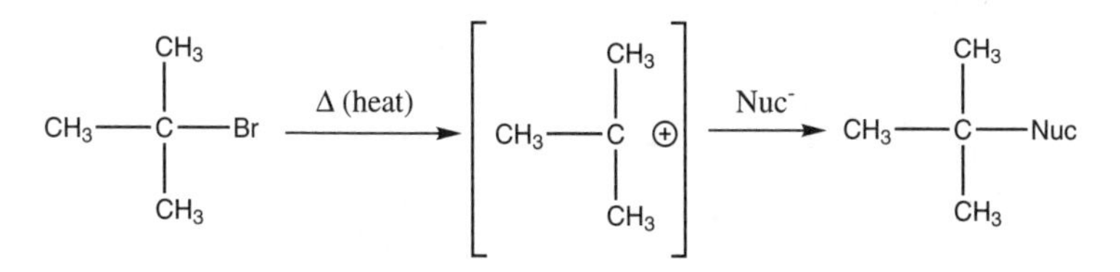

Since the carbocation is so reactive, the strength of the nucleophile has no effect on the rate of S_n1 substitution. Also since a cation must be formed, this reaction is limited to 3° and 2° halides. The success of S_n1 reactions is sensitive to the nature of the leaving group: only the better leaving groups will permit this reaction to proceed.

The mechanism of the S_n2 reaction is quite different. In this mechanism, the nucleophile attacks the alkyl halide at the same time that the bond to the leaving group is breaking. There is a requirement that the nucleophile must attack the carbon that bears the halogen directly behind the C–X bond, also known as "backside attack." As the new bond forms and the C–X bond breaks, the carbon undergoes "**inversion**," as shown in the example on the next page, in which the nucleophile is iodide and the leaving group is bromide.

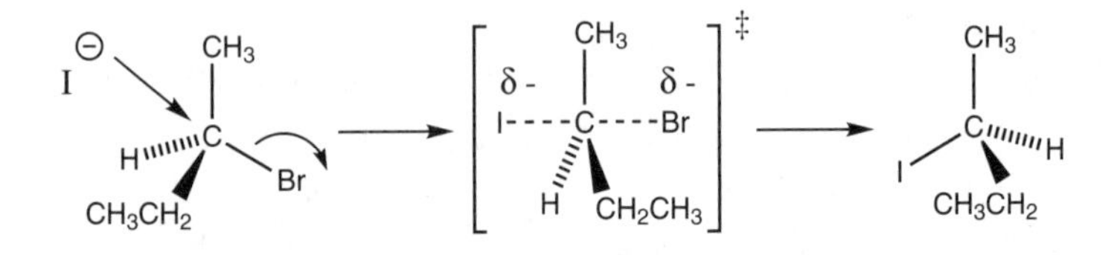

Notice that the starting material has the R configuration, but the product has the S configuration. This inversion of stereochemistry is a hallmark of the S_n2 reaction. In contrast, racemization occurs in S_n1 reactions since the sterochemistry of the starting material is destroyed as the planar cation intermediate is formed.

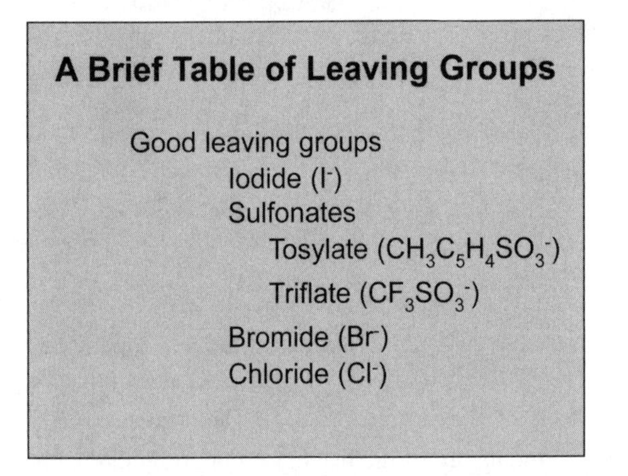

A Brief Table of Leaving Groups

Good leaving groups
- Iodide (I^-)
- Sulfonates
  - Tosylate ($CH_3C_5H_4SO_3^-$)
  - Triflate ($CF_3SO_3^-$)
- Bromide (Br^-)
- Chloride (Cl^-)

In S_n2 subsitution, the strength of the nucleophile as well as the nature of the leaving group and the substrate are all important. Only the more powerful nucleophiles will successfully react with alkyl halides. The alkyl halide substrate must not be sterically crowded, or the nucleophile will not be able to approach closely enough to displace the leaving group. As a result, S_n2 reactions are most favorable with methyl halides (such as CH_3I) and 1° halides (like CH_3CH_2Br), and never occur at 3° centers.

Elimination Reactions. Elimination reactions, which produce alkenes, can occur in competition with substitution reactions. As in substitution reactions, there are 2 common mechanisms for elimination reactions, the E1 and E2 mechanisms. In a β-elimination (dehydrohalogenation) reaction, a halogen and a hydrogen atom are removed from adjacent carbon atoms to form a double bond between the two C's. The reagent commonly used to remove HX is the strong base KOH in ethanol.

The E1 Mechanism. Like the S_n1 mechanism, the E1 mechanism is a 2-step process that proceeds via a cation intermediate.

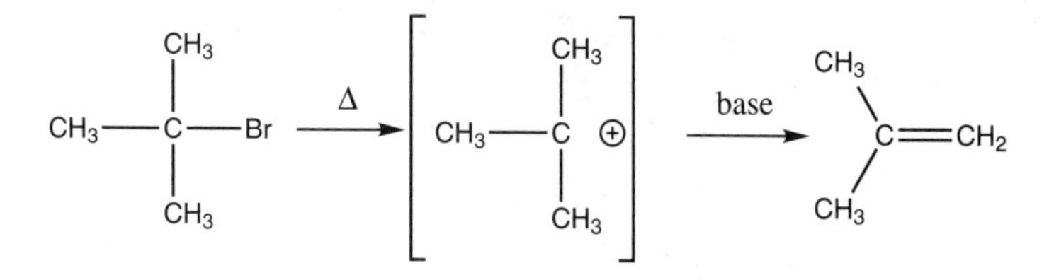

The E2 Mechanism. The E2 reaction is a single-step, bimolecular reaction in which no intermediate is formed. This reaction proceeds via a transition state that has an **antiperiplanar** arrangement of the leaving group and the proton that is being removed. In this arrangement, the hydrogen and the leaving group lie in the same plane, pointing in opposite directions. As a result, the reaction is **stereospecific**—only one of the possible cis/trans stereoisomers is formed.

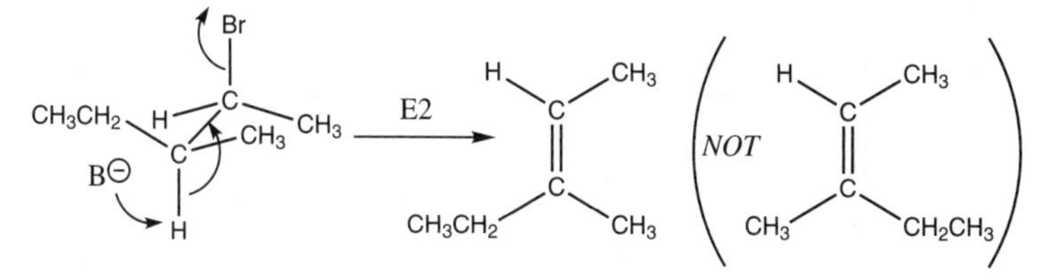

(H and Br are antiperiplanar)

Substitution versus Elimination. S_n2 reactions are preferred when the halide is a good leaving group (I^-, Br^-) and when the substrate is unhindered (methyl, 1°, or 2°) and the nucleophile is a weak base (such as Cl^-, Br^-, or I^-). E2 is preferred when a strong base is used (KOH, $NaOCH_3$). S_n1 and E1 can also compete, with the major products usually resulting from substitution.

You Need to Know

- Preparation of alkyl halides
- Nucleophiles and bases
- S_n1, S_n2, E1, E2 reactions and their mechanisms

Solved Problems

Problem 6.1 Give the organic product in the following substitution reaction.

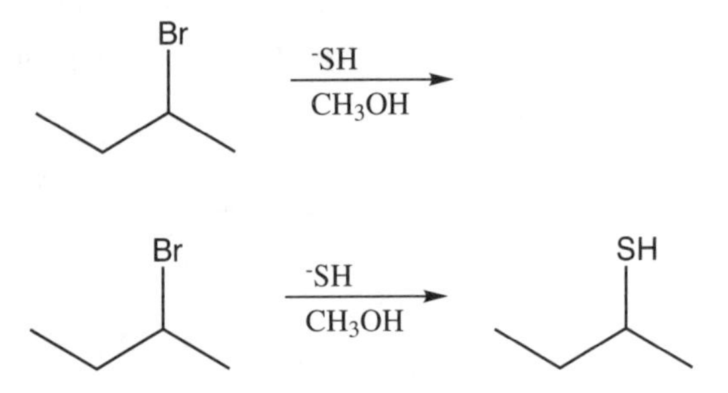

Problem 6.2 Indicate the products of the following reactions and point out the mechanism as S_N1, S_N2, E1, or E2.

(a) $(CH_3)_3CBr$ + CH_3CH_2OH, heat

(b) $CH_3CHBrCH_3$ + $NaOCH_3$ in CH_3OH

(a) $(CH_3)_3COCH_2CH_3$ (major, via S_N1)
$(CH_3)_2C{=}CH_2$ (minor, E1)

(b) $CH_3CH{=}CH_2$ (via E2)
$CH_3CH(OCH_3)CH_3$ (via S_N2)

Chapter 7
AROMATIC COMPOUNDS

IN THIS CHAPTER:

- ✔ *Introduction*
- ✔ *Electronic Structure*
- ✔ *Aromaticity and Hückel's Rule*
- ✔ *Antiaromaticity*
- ✔ *Nomenclature*
- ✔ *Chemical Reactions*
- ✔ *Order of Introducing Groups*
- ✔ *Solved Problem*

Introduction

Benzene, (C_6H_6) is the prototype of **aromatic** compounds, which are unsaturated compounds showing a low degree of reactivity. Benzene exists as a resonance hybrid of two structures shown below. As a result, all of the C–C bonds and all the hydrogens in benzene are equivalent.

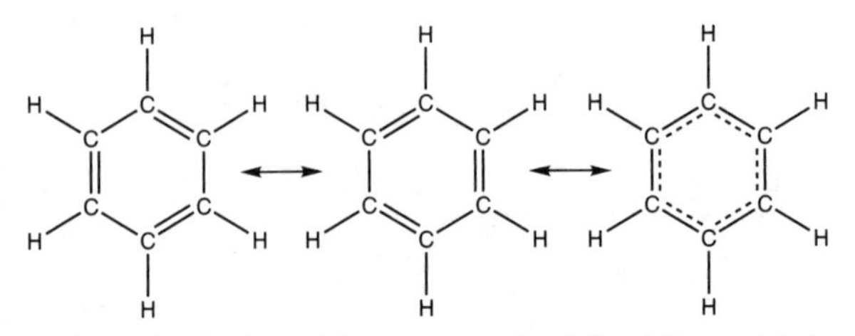

There are three disubstituted benzenes—the 1,2-, 1,3-, and 1,4-position isomers—designated as ortho, meta, and para, respectively.

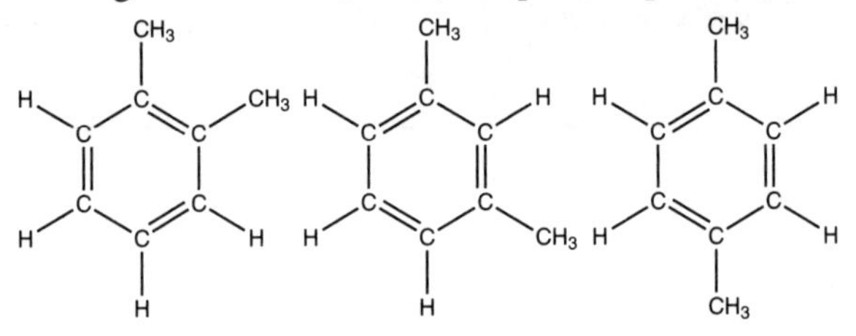

1,2- or ortho (o-) *1,3- or meta (m-)* *1,4- or para (p-)*

Electronic Structure

Each carbon atom in benzene is sp^2 hybridized. The σ bonds comprise the skeleton of the molecule. Each carbon also has a p orbital at right angles to the plane of the ring, forming a cyclic, conjugated π system containing 6 electrons. This π system is parallel to and above and below the plane of the ring.

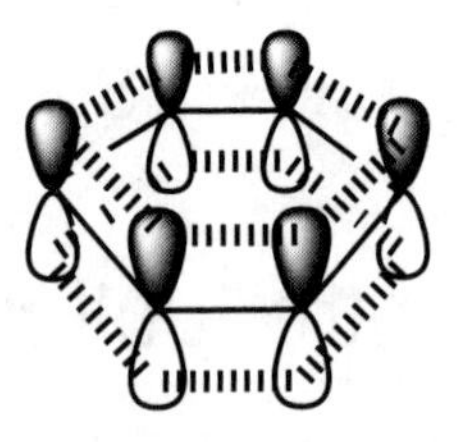

The six π electrons in the π system are associated with all six C's. They are therefore more *delocalized,* accounting for the great stability of benzene and other aromatic rings.

Aromaticity and Hückel's Rule

Hückel's rule states that if the number of π electrons in a planar, cyclic, conjugated structure is equal to $4n + 2$, where *n* equals zero or a whole number, the species is aromatic. This rule applies to heterocycles (rings containing a non-carbon atom such as nitrogen or sulfur) as well as to carbocycles (like benzene).

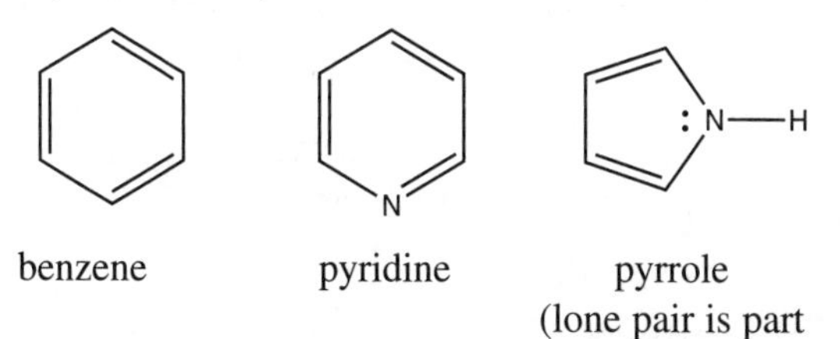

benzene pyridine pyrrole (lone pair is part of the π system)

Polycyclic Aromatic Compounds. The prototype of this class of compounds is naphthalene, $C_{10}H_8$, although there are many others. Although the Hückel 4n + 2 rule is rigorously derived for monocyclic systems, it can also be applied to other compounds.

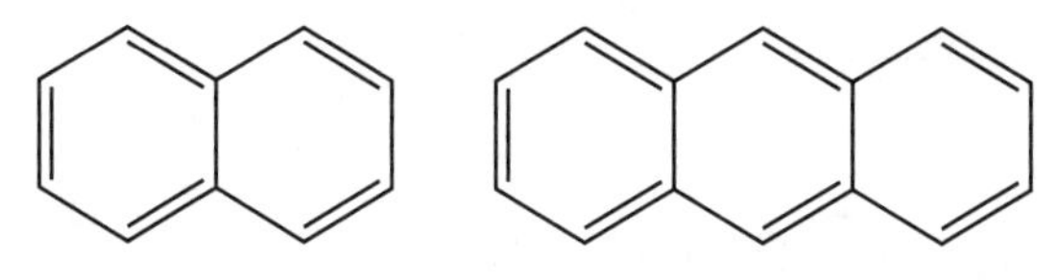

Naphthalene

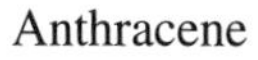

Anthracene

Antiaromaticity

Planar cyclic conjugated species with *4n* π electrons are called **antiaromatic** and are quite unstable. 1,3-Cyclobutadiene *(n* = 1*)*, for which two equivalent contributing resonance structures can be writtten, is an extremely unstable antiaromatic molecule.

Nomenclature

Some benzene derivatives are usually referred to by their common, nonsystematic names, such as toluene ($C_6H_5CH_3$), xylene ($C_6H_4(CH_3)_2$), phenol (C_6H_5OH), and aniline ($C_6H_5NH_2$). **Derived names** combine the name of the substituent as a prefix with the word *benzene*. Examples include nitrobenzene ($C_6H_5NO_2$) and ethylbenzene ($C_6H_5CH_2CH_3$). Some common aromatic groups that are substituents (Ar–) are C_6H_5– (phenyl), C_6H_5–C_6H_4– (biphenyl), and p–$CH_3C_6H_4$(p-tolyl). Another common group is $C_6H_5CH_2$–, known as benzyl.

For disubstituted benzenes with a group giving the ring a common name, o-, p-, or m- is used to designate the position of the second group. Otherwise, positions of groups are designated by the lowest combination of numbers.

Chemical Reactions

The unusual stability of the benzene ring dominates the chemical reactions of benzene and naphthalene. Both compounds resist addition reactions which lead to destruction of the aromatic ring. Rather, they undergo substitution reactions, in which a group or atom replaces an H from the ring, thereby preserving the stable aromatic ring. Atoms or groups other than H may also be replaced.

Reduction. Benzene is resistant to catalytic hydrogenation (high temperatures and high pressures of H_2 are needed) and to reduction with Na in alcohol. Reduction with lithium in liquid ammonia (known as the **Birch** reduction) produces 1,4-cyclohexadiene.

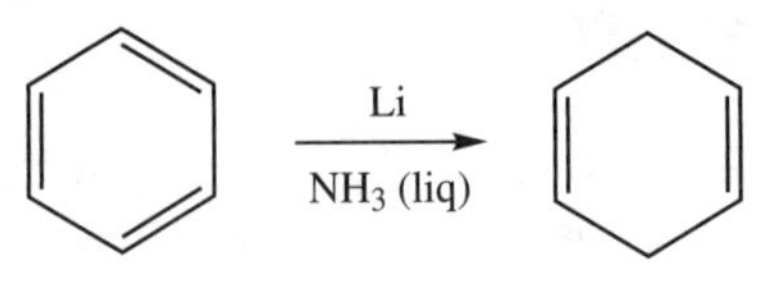

Oxidation. Benzene is very stable to oxidation except under very vigorous conditions. In fact, when an alkylbenzene is oxidized, the alkyl group is oxidized to a COOH group, while the benzene ring remains intact. For this reaction to proceed, there must be at least one H atom on the C attached to the ring.

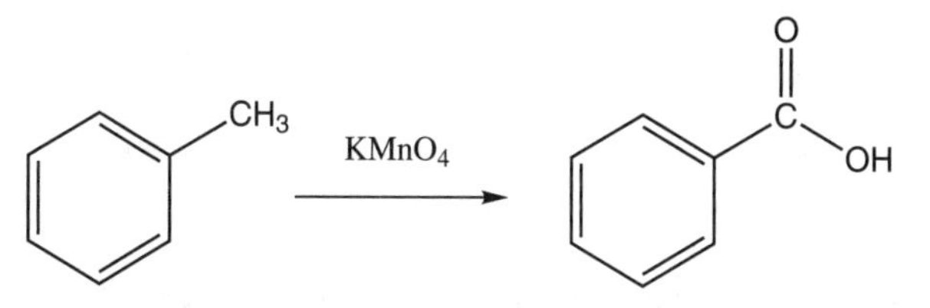

Electrophilic Aromatic Substitution. Aromatic rings undergo substitution reactions (replacement of a hydrogen with another group) with strong electrophiles. The mechanism for these reactions involve attack of the electrophile on the π electrons of the ring, followed by loss of a proton to reestablish aromaticity. These reactions typically require Lewis acid catalysts to help generate the electrophiles. The first step in this reaction is reminiscent of electrophilic addition to an alkene. Aromatic substitution differs in that the intermediate carbocation loses a cation (most often H^+) to give the substitution product, rather than adding a nucleophile to give the addition product.

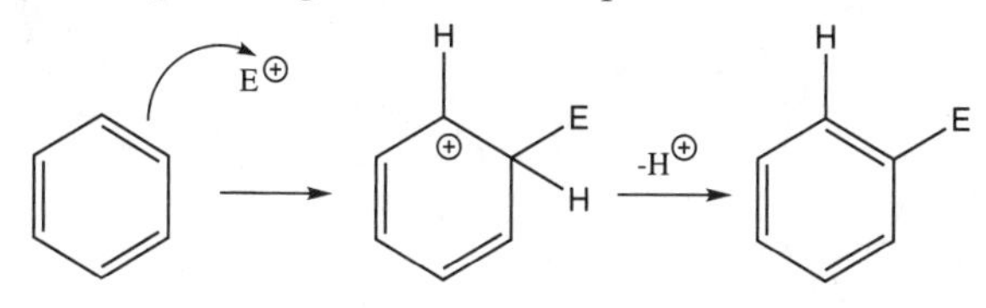

A wide range of different electrophiles can be used, as demonstrated below.

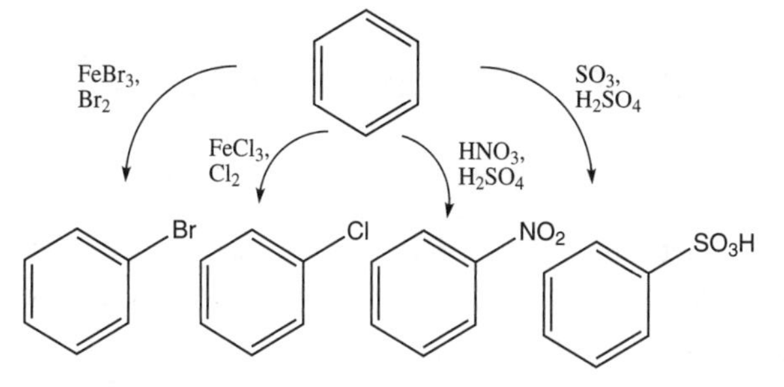

The introduction of acyl groups is accomplished by treating benzene with an acid chloride and $AlCl_3$. This reaction is known as the **Friedel-Crafts Acylation**.

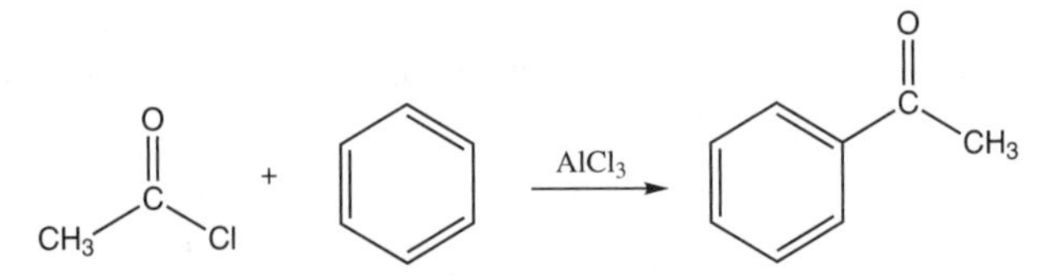

The 5 ring H's of monosubstituted benzenes are not equally reactive. The ring substituents determine the orientation of E *(meta*, or a mixture of *ortho* and *para),* and the reactivity of the ring toward substitution. Electron-donating groups (–OCH_3, –NR_2, alkyl) make the ring more reactive, and direct electrophilic attack to the *ortho* and *para* positions. These groups are known as **ortho, para directors**. Electron withdrawing groups (–NO_2, acyl) are **meta directors** and deactivate the ring toward electrophilic attack. Halogens are ortho, para directors, but weak deactivators.

You Need to Know

- Aromaticity, antiaromaticity, and Hückel's Rule
- Electrophilic aromatic substitution
- o/p- and m-directors

Order of Introducing Groups

In the preparation of highly substituted aromatic compounds, it is essential to introduce the substituents in the proper order. For example, if benzene is first treated with CH_3Cl in the presence of $AlCl_3$, then with HNO_3/H_2SO_4 the para product will be formed. Reversing the order of these steps yields the meta product.

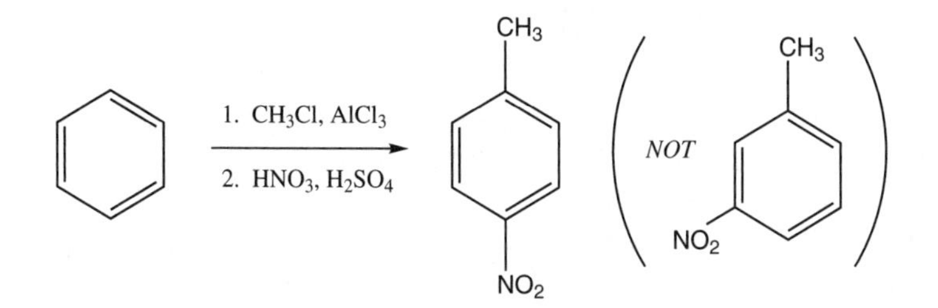

Reactivity of the Benzylic Positions

Benzylic carbons are adjacent to an aromatic ring. The chemistry of benzylic and of allylic positions are very similar. Intermediate carbocations, free radicals, and carbanions formed at these positions are stabilized by delocalization with the adjacent π system, the benzene ring in the case of the benzylic position. Benzylic halides can be prepared in good yield through free-radical halogenation, as shown below.

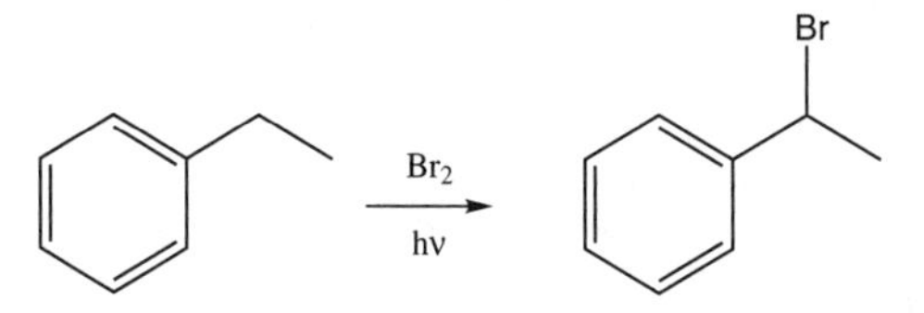

Benzylic halides are highly reactive, even reacting with nucleophiles as weak as water. Alkyl halides do not undergo nucleophilic substitutions with such weak nucleophiles.

Solved Problem

Problem 7.1 Use PhH, PhMe, and any aliphatic or inorganic reagents to prepare the following compounds in reasonable yields.
(a) m-bromobenzenesulfonic acid
(b) 3-nitro-4-bromobenzoic acid

(a)

SO_3H SO_3H

SO_3 / H_2SO_4 → ; Br_2 / $FeBr_3$ → ; Br

(m-director is added first)

(b)

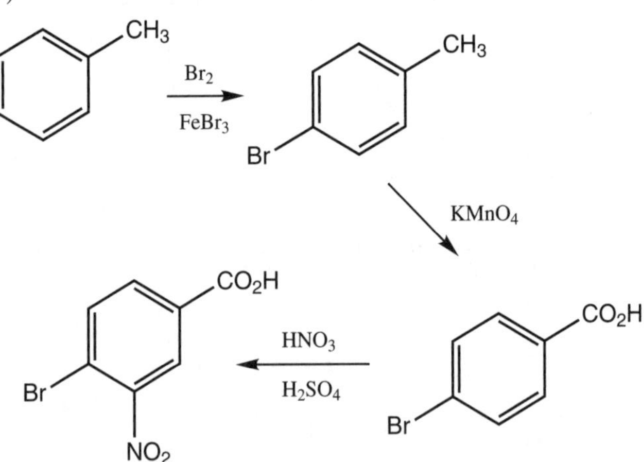

Nitration of p-$BrC_6H_4CH_3$ would have given about a 50-50 mixture of two products; 2-nitro-4-bromotoluene would be unwanted. When oxidation precedes nitration, an excellent yield of the desired product is obtained.

Chapter 8
SPECTROSCOPY AND STRUCTURE

IN THIS CHAPTER:

- ✔ *Introduction*
- ✔ *Ultraviolet-Visible Spectroscopy*
- ✔ *Infrared Spectroscopy*
- ✔ *Nuclear Magnetic Resonance*
- ✔ *Mass Spectrometry*
- ✔ *Solved Problems*

Introduction

Spectral properties are used to determine the structure of molecules and ions. Of special importance are ultraviolet (UV), infrared (IR), nuclear magnetic resonance (NMR), and mass spectra (MS).

Different types of electromagnetic radiation correspond to different molecular transformations depending on the energy separation between the two states of the molecule involved. The molecule can be raised from its lowest energy state **(ground state)** to a higher energy state **(excited state)** by a photon (quantum of energy) of electromagnetic radiation of the correct wavelength.

Region of Electromagnetic Spectrum	Type of Excitation
Ultraviolet (UV)	Electronic
Visible (Vis)	Electronic
Infrared (IR)	Molecular vibration
Radio	Spin (electronic or nuclear)

Wavelengths (λ) for UV-visible spectra are typically expressed in **nanometers** (1 nm = 10^{-9} m), while for infrared spectra, **wavenumbers** (reciprocal centimeters, or cm^{-1}) are used.

In a typical spectrophotometer, a dissolved compound is exposed to electromagnetic radiation over a specified wavelength. The radiation passing through the sample is recorded as a function of the wavelength or wavenumber.

Ultraviolet-Visible Spectroscopy

Ultraviolet or visible light cause an electron to be excited from a lower-energy orbital to a high-energy orbital. Compounds that absorb this type of radiation are unsaturated, and conjugated double bonds absorb lower energy (longer wavelengths in the visible region) than isolated double bonds do. Species that absorb in the visible region are colored, and black is observed when all visible light is absorbed.

Infrared Spectroscopy

Infrared radiation causes excitation of molecular vibration states. Diatomic molecules such as H–H and H–Cl vibrate in only one way; the atoms move, as though attached by a coiled spring, toward and away from each other. This mode is called **bond stretching.** Triatomic molecules, such as CO_2 (O=C=O), possess two different stretching modes. In the **symmetrical stretch** mode, each O moves away from the C at the same time. In the **antisymmetrical stretch,** one O moves toward the C while the other O moves away.

An observed absorption band at a specific wavelength indicates the presence of a particular bond or group of bonds in a molecule. Although absorption is affected only slightly by the molecular environment of the bond or group, it is possible to determine from small variations in the band frequencies such factors as the size of the ring containing a C=O group or whether C=O is part of a ketone, acid chloride, or an acid. Conversely, the absence of a certain band in the spectrum usually rules out the presence of the bond that would produce it.

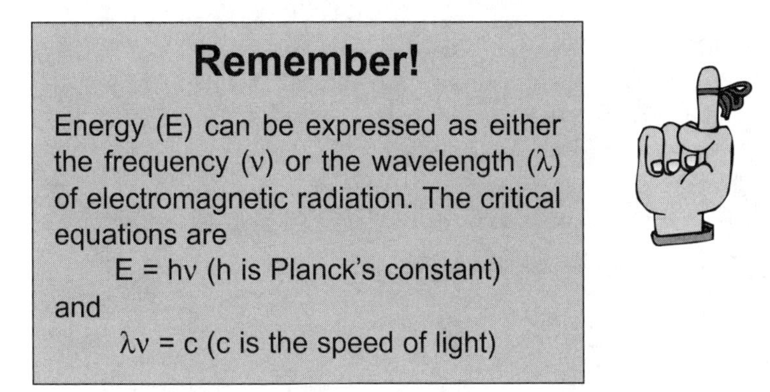

Between 1400 and 800 cm^{-1}, there are many peaks which are difficult to interpret. This range, called the **fingerprint region,** is useful for determining whether two samples are composed of the same compound. No two compounds have exactly same IR spectrum.

The C–H stretch absorption, the C=O stretch absorption, and the fingerprint region are all visible in the IR spectrum of ethyl acetate, shown below.

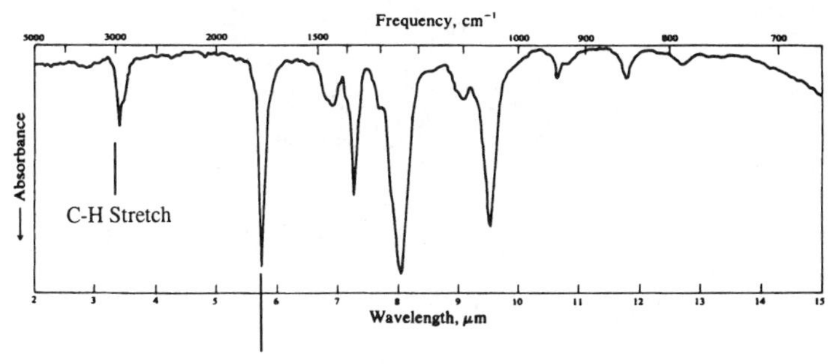

You Need to Know

Infrared Absorption Peaks

cm^{-1}	Intensity	Structure
1340,1500	(s)	NO_2
1450–1600	(s)	C=C bond in aromatic ring
1620–1680	(w)	C=C
1630–1690	(s)	C=O (in amides)
1690–1750	(s)	C=O (in carbonyl compounds and esters)
1700–1725	(s)	C=O (in carboxylic acids)
1770–1820	(s)	C=O (in acid chlorides)
2100–2200	(w)	C≡C
2210–2260	(m)	C≡N
2700–2800	(w)	C=H (of aldehyde group)
2500–3000	(s, b)	O–H in COOH
3000–3100	(m)	C–H (C is part of aromatic ring)
3300	(s)	C–H (C is acetylenic)
3020–3080	(m)	C–H (C is alkene or aromatic ring)
2800–3000	(ms)	C–H (in alkanes)
3300–3500	(m)	N–H (in amines and amides)
3200–3600	(s, b)	O–H (in H-bonded ROH and ArOH)
3600–3650	(s)	O–H

*Intensities: (s) = strong, (m) = medium, (w) = weak, (b) = broad

^{1}H Nuclear Magnetic Resonance

Nuclei with an odd number of protons or neutrons have permanent magnetic moments and quantized nuclear spin states. When a compound is placed in a magnetic field, its H's align their own fields either with or against the applied magnetic field, H_0, giving rise to two separated energy states. The lower energy state is the one with the hydrogen's magnetic field lined up with the external field.

The difference in energy between the two states is directly proportional to H_0. Radio-frequency radiation can "flip" hydrogen nuclei from lower to higher energy states. When this spin-flip occurs, the nucleus (a proton, in the case of hydrogen) is said to be in **resonance** with the applied radiation, leading to the name **nuclear magnetic resonance (NMR)** spectroscopy. When used for medical diagnosis this technique is called **magnetic resonance imaging (MRI).** The excited nuclei quickly return to the lower-energy spin state.

Chemical Shift. NMR spectroscopy is useful because, in a given magnetic field, not all H's change spin at the same radio frequency. The magnetic field experienced by a hydrogen is not necessarily that which is applied by the magnet, because the electrons in the bond to the hydrogen and the electrons in nearby bonds *induce* their own magnetic fields. This induced field, H^*, partially shields the proton from the applied H_0. The field "felt" by the proton, the **effective field**, is $H_0 - H^*$.

For historical reasons, resonances (peaks) on the left side of the spectrum are called *downfield*, and resonances on the right side of the spectrum are called *upfield.*

The chemical shift refers to how much a peak is shifted from the resonance of a reference compound, tetramethylsilane. This compound serves as a useful reference because its ^{1}H resonance is usually isolated in the extreme upfield region of the spectrum.

Electronegative atoms, such as N, O, and X (halogen), lessen the shielding of H's and cause downfield shifts. The extent of the downfield shift is directly proportional to the electronegativity of the atom and its proximity to the H. The influence of electronegativity is illustrated with the methyl halides, MeX.

Compound	δ, ppm
CH_3F	4.3
CH_3Cl	3.1
CH_3Br	2.7
CH_3I	2.2

Hydrogens attached to π-bonded C's are less shielded than those in alkanes are. Ar, C=O, and C=C are electron withdrawing by induction, causing a downfield shift of an H on an adjacent C.

For alkyl groups in similar environments, shielding increases with the number of H's on the C. Thus, CH_3 groups tend to be farther upfield than CH_2 groups or CH groups.

Hydrogens which participate in H bonding, e.g., OH and NH, exhibit variable δ values over a wide range. Hydrogen bonding is accompanied by exchange of hydrogens between the ROH molecules, resulting in broadening of the signals.

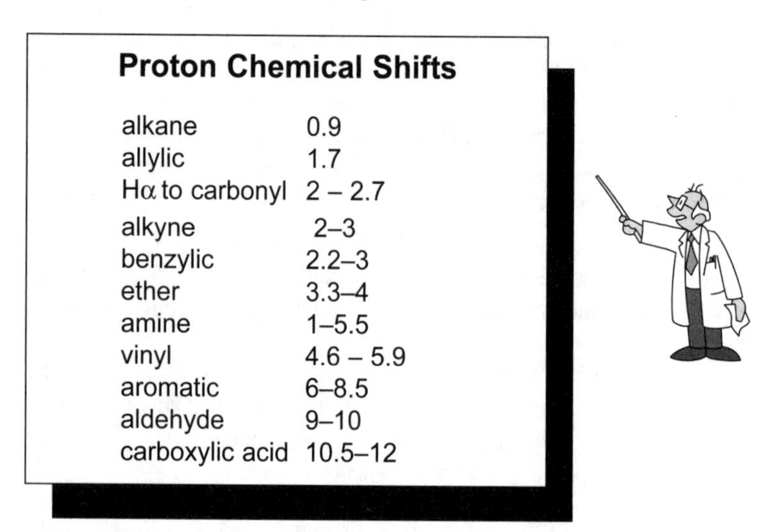

Proton Chemical Shifts

alkane	0.9
allylic	1.7
Hα to carbonyl	2 – 2.7
alkyne	2–3
benzylic	2.2–3
ether	3.3–4
amine	1–5.5
vinyl	4.6 – 5.9
aromatic	6–8.5
aldehyde	9–10
carboxylic acid	10.5–12

Integrals. The area under a peak is directly proportional to the number of equivalent H's giving the signal. For example, the compound $C_6H_5CH_2C(CH_3)_3$ has five aromatic protons (a), two benzylic protons (b), and nine equivalent CH_3 protons (c). Its NMR spectrum shows three peaks for the three different kinds of H, which appear at: (a) 7.1 ppm (aromatic H), *(b)* 2.2 ppm (benzylic H), and (c) 0. 9 ppm (t-butyl H). The relative areas under the peaks a : b : c equals 5 : 2 : 9.

Peak Splitting: Spin-Spin Coupling. Because of **spin-spin coupling,** most NMR spectra do not show simple single peaks but rather groups of peaks that tend to cluster about certain values. To see how this coupling arises, we examine a molecular fragment present in a very large number of molecules.

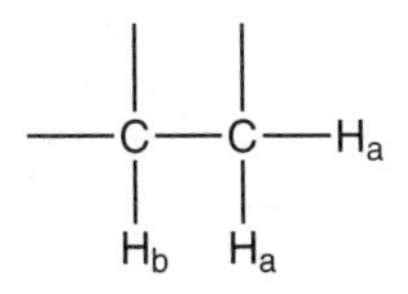

The signal for H_a is shifted slightly upfield or downfield depending on whether the spin of H_b is aligned against or with the applied field. Since in about half of the molecules the H_b are oriented up and in half down, H_a appears as a doublet instead of a singlet. The effect is reciprocal: the two H_a's also split the signal of H_b. There are four possible states with approximately equal probability for the two H_a's.

Because the middle two spin states have the same effect, the signal of H_b is split into a triplet with relative intensities 1 : 2 : 1.

This phenomenon is demonstrated graphically below. There is only one way both H_a spins can add together to make a higher field, 2 ways they can have no net effect on the field at H_b, and one way they can add together to decrease the field felt by H_b. Therefore the H_b resonance is split into 3 lines.

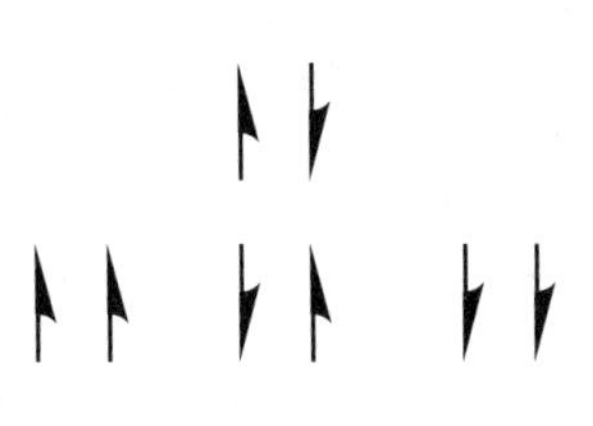

Spin-spin coupling usually occurs between *nonequivalent* H's on adjacent atoms. In general, if n equivalent H's are affecting the peak of H's on an adjacent C, the peak is split into n + 1 peaks.

The ^1H NMR spectrum of diisopropyl ether is shown below. The chemical shifts of the 2 different types of protons (CH and CH_3), the relative integrals (1:6), and the $^1H - {}^1H$ coupling are evident.

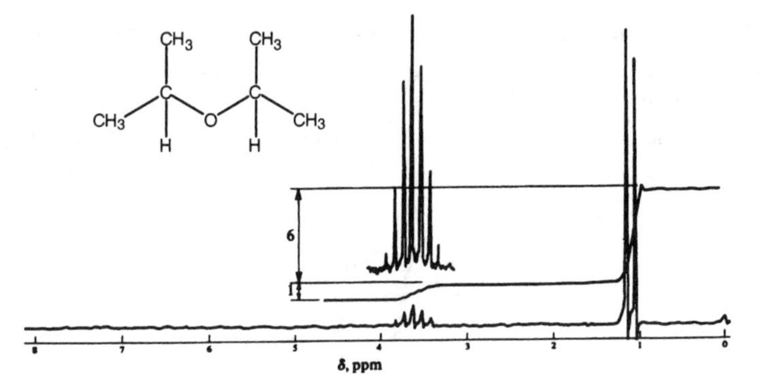

^{13}C Nuclear Magnetic Resonance

Although ^{12}C is not NMR-active, the ^{13}C isotope, which has a natural abundance of about 1%, is NMR-active. Since ^1H and ^{13}C absorb at different frequencies, proton signals do not appear in the ^{13}C spectra. However, spin-spin coupling between attached H's and a ^{13}C is observed. The same $(n + 1)$-rule that is used for H's on adjacent C's is used to analyze the $^{13}C - {}^1H$ coupling pattern. To obtain simpler spectra, these samples are usually analyzed under conditions (decoupling) which prevent the protons from causing splitting. Thus, each different carbon appears as a sharp singlet. Both types of spectra are useful. The decoupled spectrum permits the counting of different carbons in the molecule; the coupled spectrum allows the determination of the number of H's attached to each carbon.

Each ^{13}C has a characteristic chemical shift δ. Note that the carbonyl carbon is the most downfield, with aromatic carbons somewhat more upfield. Carbon NMR spectra are not usually integrated, because the techniques used to decouple the peaks change the peak areas.

Mass Spectrometry

When exposed to sufficient energy, a molecule may lose an electron to form a cation radical, which then may fragment. These processes make **mass spectrometry** a useful tool for structure proof. A beam of energetic electrons ionizes parent molecules in the vapor state, and a number of **parent ions** may then fragment to give other cations and neutral species. Fragment ions can undergo further bond cleavage to give even smaller cations and neutral species. In a mass spectrum, sharp peaks appear at the values of *m/e* (the mass-to-charge ratio) for the various cations. The relative heights of the peaks represent relative abundances of the cations. Most cations have a charge of +1 and, therefore, most peaks record the masses of the cations.

The standard against which all peak intensities in a given mass spectrum are measured is the most intense peak, called the **base peak,** which is arbitrarily assigned the value 100. If few parent molecules fragment—not a typical situation—the parent cation will furnish the base peak.

Unless all the parent ions fragment, the largest observed *m/e* value is the molecular weight of the parent molecule. This generalization overlooks the presence of naturally occurring isotopes in the parent. Thus, the chances of finding a ^{13}C atom in an organic molecule are 1.11% times the number of carbons in the molecule; the chances of finding two are negligible for small and medium sized molecules. Therefore, the instrument detects a small peak at $m + 1$, due to the ^{13}C–containing parent.

The masses and possible structures of fragment ions are clues to the structure of the original molecule. Mass spectra, like other spectra, are unique properties used to identify compounds.

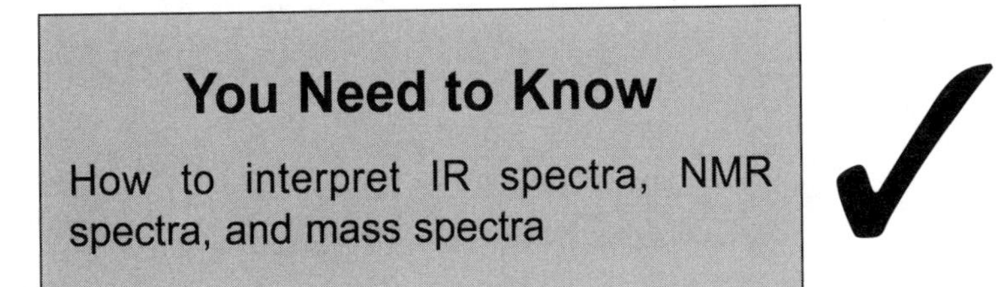

Solved Problems

Problem 8.1 Which peaks in the IR spectra distinguish cyclohexane from cyclohexene?

One of the C-H stretches in cyclohexane is above 3000 cm^{-1} (Csp^2-H); in cyclohexane the C-H stretches are below 3000 cm^{-1} (Csp^3–H). Cyclohexene has a C=C stretch at about 1650 cm^{-1}.

Problem 8.2 How can NMR distinguish between p-xylene and ethylbenzene?

p-$CH_3C_6H_4CH_3$ has two kinds of H's and its NMR spectrum has two singlets. $C_6H_5CH_2CH_3$ has three kinds of H's, and its NMR spectrum has a singlet for the aromatic hydrogens, a quartet for CH_2 H's, and a triplet for CH_3 H's.

Chapter 9 ALCOHOLS, ETHERS, AND EPOXIDES

IN THIS CHAPTER:

- ✔ *Nomenclature and H Bonding in Alcohols*
- ✔ *Preparation of Alcohols*
- ✔ *Reactions of Alcohols*
- ✔ *Ethers*
- ✔ *Epoxides*
- ✔ *Solved Problems*

Nomenclature and H Bonding in Alcohols

ROH is **an alcohol** and ArOH is a **phenol.** Some alcohols have common names, usually made up of the name of the alkyl group attached to the OH and the word "alcohol"; e.g., *ethyl alcohol,* CH_3CH_2OH. In the IUPAC method the suffix -ol replaces the -e of the alkane to indicate the OH. The longest chain with the OH group is used as the parent. Under this system, CH_3CH_2OH is called ethanol.

Alcohols boil at a higher temperature than the corresponding hydrocarbons, due to the hydrogen bonding between molecules in the liquid. This tendency to form hydrogen bonds causes small molecular weight alcohols (four or fewer carbons) to be water-soluble. As the R group becomes larger, ROH resembles the hydrocarbon more closely.

Preparation of Alcohols

By S_N2 or S_N1 from alkyl halides. Alcohols can be prepared from alkyl halides using hydroxide or water as the nucleophile. The mechanism of the reaction depends on the alkyl group. Primary alkyl halides undergo S_N2 reaction with hydroxide. With secondary and tertiary halides, competing elimination reactions with hydroxide are a problem. Using water as the nucleophile can minimize these side reactions.

Hydration of Alkenes

Alcohols can be prepared by adding water across the C=C bond of alkenes. Acid-catalyzed hydration follows Markovnikov's rule and is often accompanied by side reactions or rearrangements.

Oxymercuration-demercuration of alkenes leads to net addition of H–OH, giving a product that follows Markovnikov's rule and is free from rearrangement.

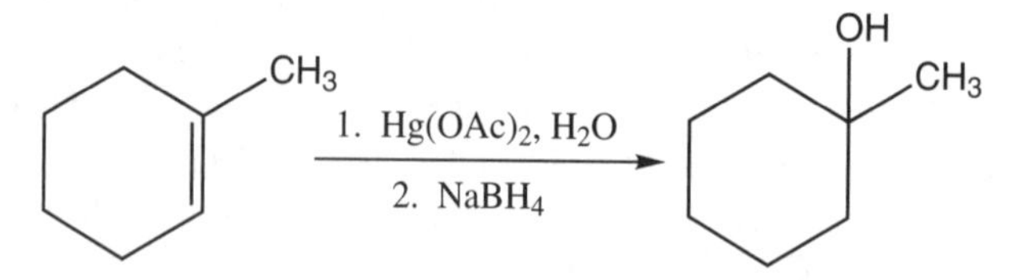

Anti-Markovnikov addition of water can be carried out via **hydroboration-oxidation** of alkenes. Treatment of alkylboranes with H_2O_2 in ^-OH replaces B with OH. The net addition of H–OH to alkenes is cis, anti-Markovnikov, and free from rearrangement.

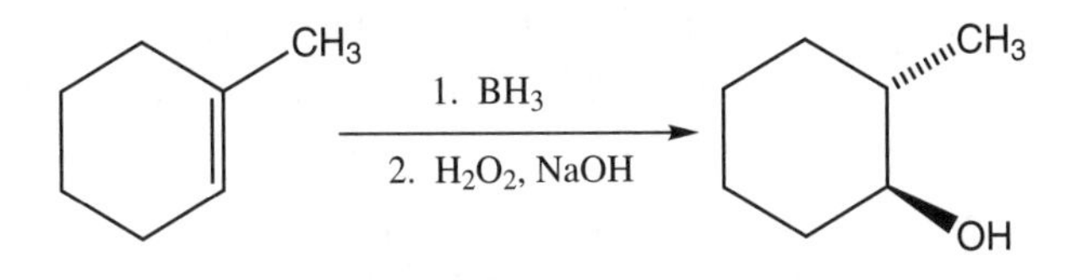

Carbonyl Compounds and Grignard Reagents. Grignard reagents, RMgX and ArMgX, are reacted with aldehydes or ketones and the intermediate salts are then hydrolyzed to alcohols. The nucleophilic Grignard reagent attacks the partially positive carbon of the carbonyl group.

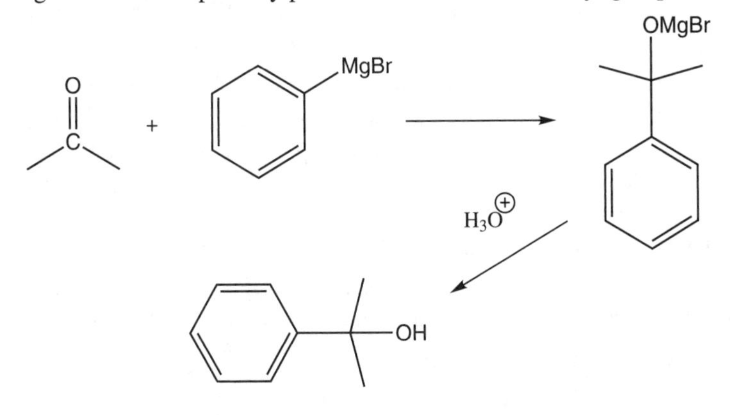

Primary alcohols can be made by adding a Grignard reagent to formaldehyde ($H_2C=O$). Secondary aldohols are made by combining Grignard reagents with an aldehyde, and tertiary alcohols are prepared by the reaction of a Grignard reagent with a ketone.

There are several limitations of Grignard reagents, due to functional groups that react with this reagent. The halide cannot possess a functional group with an acidic H, such as OH, CO_2H, NH, or SH, because then the carbanion of the Grignard group would remove the acidic H.

Carbonyl groups in the Grignard reagent will lead it to react inter- or intramolecularly with itself.

Reduction of Carbonyl Compounds. Alcohols can be readily formed by reacting aldehydes or ketones with sodium borohydride, $NaBH_4$, in protic solvents such as ROH or H_2O. Lithium aluminum hydride, $LiAlH_4$, in anhydrous ether can also be used. Esters and carboxylic acids can be reduced to alcohols using $LiAlH_4$. The initial product is a lithium alkoxide salt, which is then hydrolyzed to a 1° alcohol by addition of dilute acid. The alkoxy part of the ester is cleaved off in this reduction.

Reactions of Alcohols

The H on an alcohol OH group is weakly acidic. Sodium hydride (NaH), sodium amide ($NaNH_2$), organolithium reagents (RLi), and Grignard reagents (RMgX) are basic enough to deprotonate the alcohol group.

Alcohols react with HX to form alkyl halides. This reaction proceeds rapidly for tertiary alcohols. Primary and secondary alcohols react more slowly, requiring prolonged heating, which often leads to side reactions. These side reactions can be avoided by using thionyl chloride ($SOCl_2$) or phosphorus tribromide (PBr_3) to make chlorides or bromides. Both of these reagents convert the OH into a better leaving group, allowing halide to act as a nucleophile in an S_N2 reaction. The p-toluenesulfonate anion, (tosylate, p-$CH_3C_6H_5SO_3^-$) is a very weak base and is therefore a good leaving group.

> Alcohols are readily converted to tosyl esters (tosylates) by reacting them with tosyl chloride (p-$CH_3C_6H_4SO_2Cl$). The tosyl ester is much more reactive to nucleophilic substitution than the precursor alcohol.

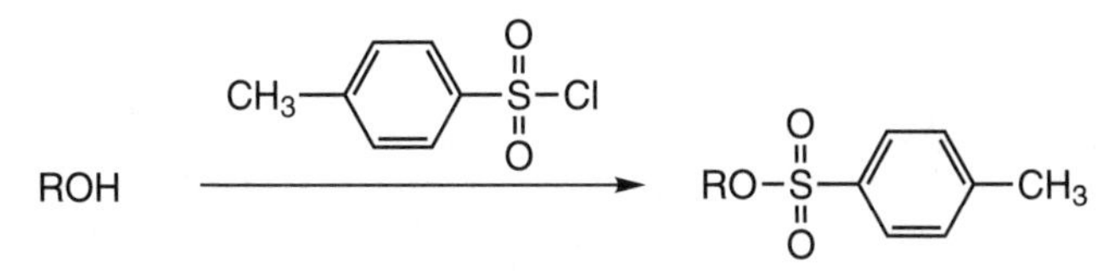

Upon heating alcohols with non-nucleophilic acids (H_2SO_4 and H_3PO_4), dehydration to alkenes occurs. Intermolecular dehydrations can also occur, leading to ethers, which are discussed in a later part of this chapter.

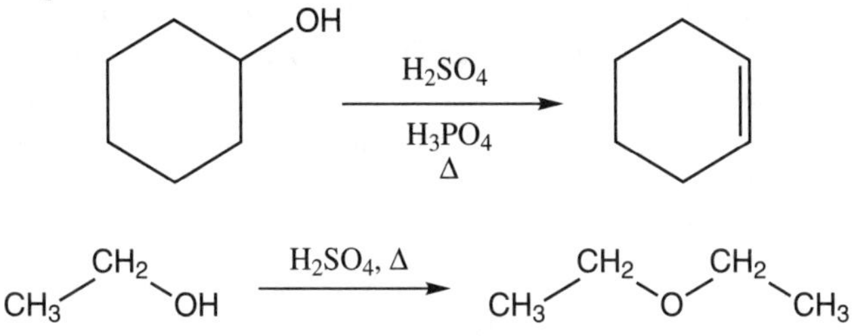

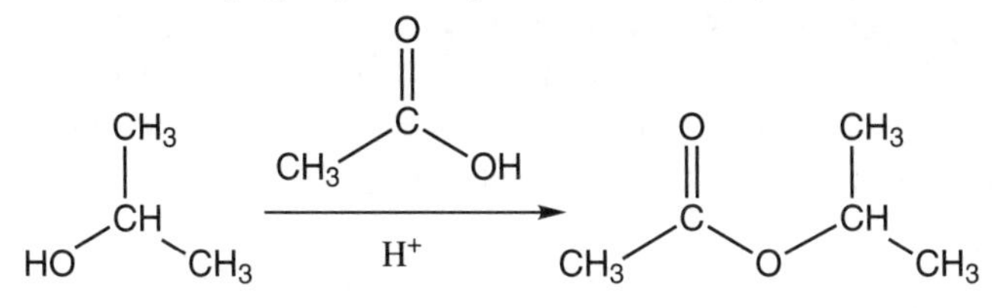

In the presence of an acid catalyst, alcohols react with carboxylic acids to yield esters. In these reactions, the alcohol acts as a nucleophile and attacks the carbonyl group of the protonated carboxylic acid.

$$\mathrm{HO{-}CH(CH_3){-}CH_3} \xrightarrow[\mathrm{H^+}]{\mathrm{CH_3C(=O)OH}} \mathrm{CH_3C(=O){-}O{-}CH(CH_3){-}CH_3}$$

Alcohols with at least one H on the alcohol carbon are oxidized with Jones reagent (CrO_3, H_2SO_4) to carbonyl compounds. Primary alcohols yield aldehydes, which are further oxidized to carboxylic acids, RCO_2H. To obtain aldehydes selectively, milder reagents, such as pyridinium chlorochromate (PCC), are used.

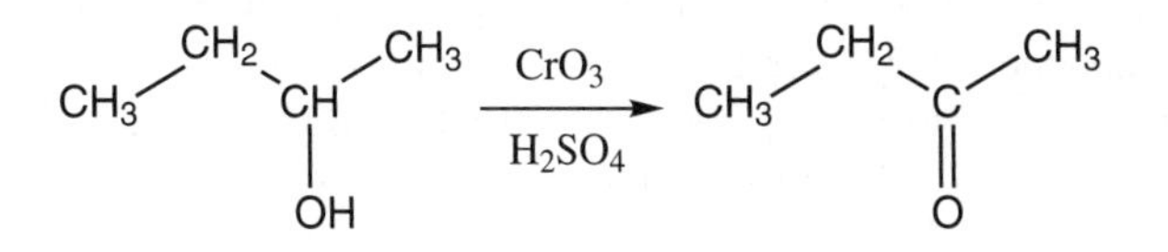

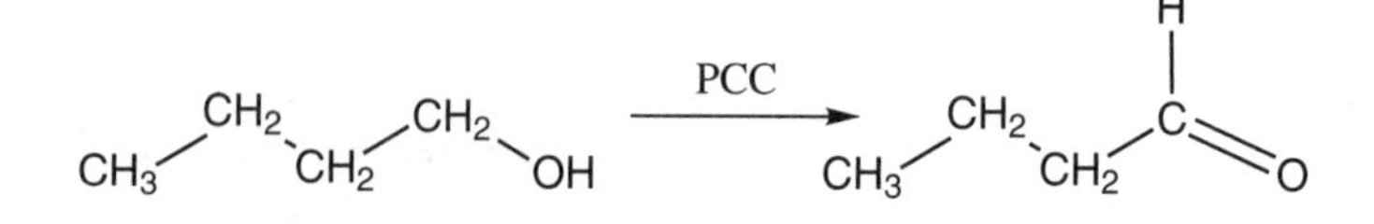

Ethers

Simple (symmetrical) ethers have the general formula ROR or ArOAr; mixed (unsymmetrical) ethers are ROR', ArOAr' or ArOR. The common system names R and Ar as separate words and adds the word "ether." In the IUPAC system, ethers (ROR) are named as alkoxy- (RO–) substituted alkanes.

methyl ethyl ether
or methoxy ethane

Symmetrical ethers can be prepared by intermolecular dehydration of alcohols in the presence of acid. Unsymmetrical ethers can be prepared via the **Williamson Synthesis**. In this reaction, an alkoxide acts as a nucleophile with an alkyl halide. Due to the competing elimination reactions with secondary or tertiary halides, the best yields are obtained with methyl halides or primary alkyl halides.

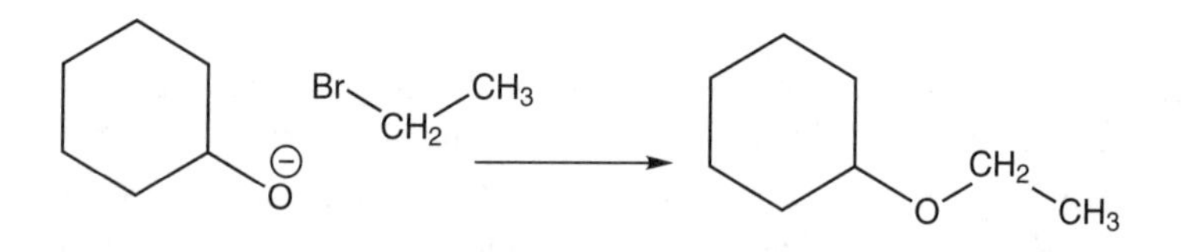

Ethers are among the most unreactive functional groups in organic chemistry. This property, together with their ability to dissolve nonpolar compounds, makes them good solvents for organic compounds.

Ethers are cleaved by concentrated HI (ROR + HI yields ROH + RI). With excess HI, 2 mol of RI are formed.

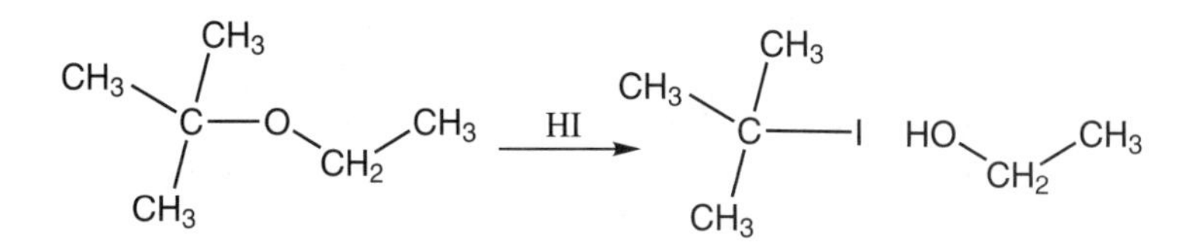

Epoxides

Unique among cyclic ethers are those with three-membered rings, the **epoxides.** *Their large ring strains make them highly reactive.* Epoxides are readily synthesized by treating alkenes with a peroxyacid (e.g., m-chloroperoxybenzoic acid). The stereochemistry of the alkene is retained in the epoxide. The reaction is a stereospecific cis addition.

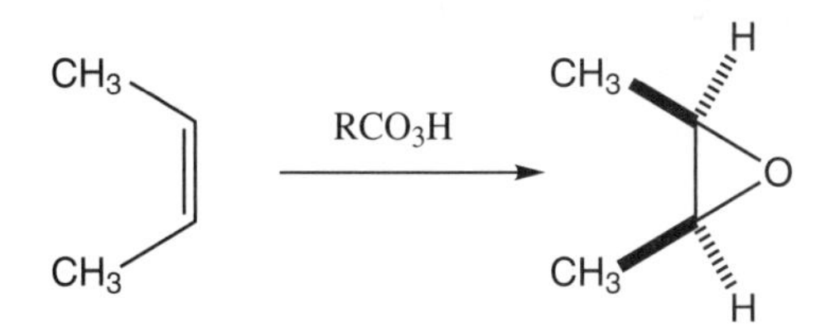

Halohydrins, formed by electrophilic addition of X_2 and H_2O to alkenes, are treated with base to give epoxides. This reaction is an intramolecular S_N2-reaction of the alkoxide on the alkyl halide.

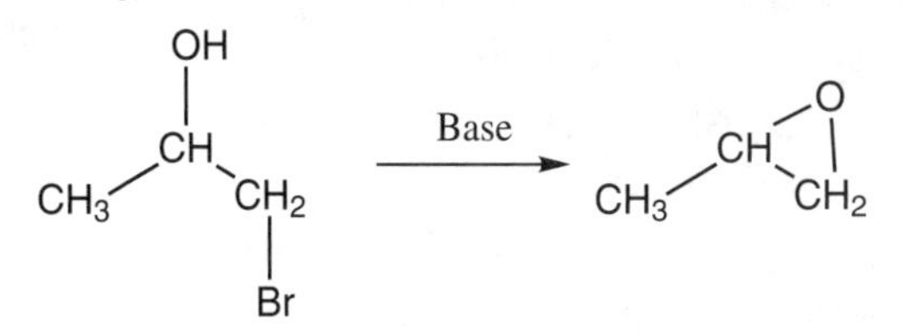

Ring-opening reactivity is attributed to the highly strained three-membered ring, which is readily cleaved. In acid the S_N1 mechanism produces the more stable (more substituted) carbocation and the nucleophilic solvent forms a bond with the more substituted C. Acid-induced ring openings are readily achieved by attack of nucleophiles, because the very weakly basic OH, formed by protonation of the O atom, is displaced as part of the alcohol portion of the product.

In base, the ring is cleaved by attack of the nucleophile on the less substituted C to form an alkoxide anion, which is then protonated. Base-induced ring-openings require a strong base, because the strongly basic RO^- is displaced as part of the alkoxide.

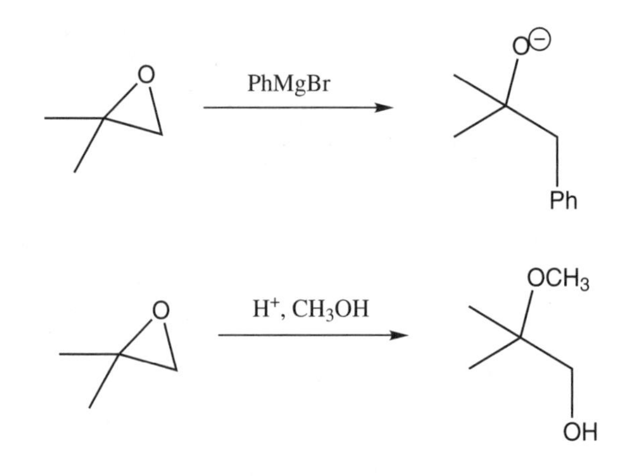

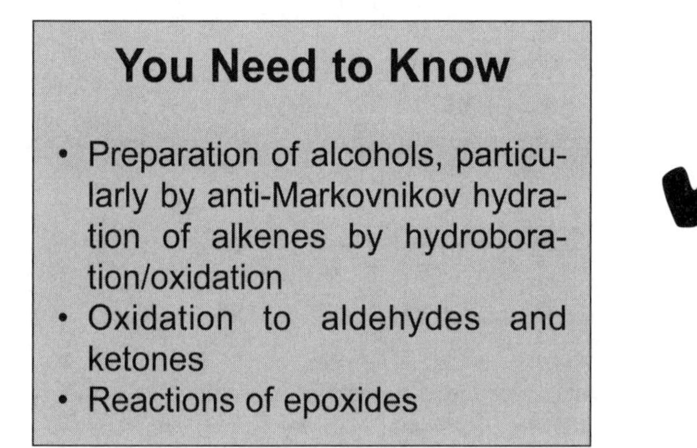

Solved Problems

Problem 9.1 Identify the products from reaction of $(CH_3)_2C{=}O$ with
(a) $NaBD_4$ followed by H_2O
(b) $NaBD_4$ followed by CH_3OD.

(a) $(CH_3)_2CDOH$
(b) $(CH_3)_2CDOD$.

In both reactions, nucleophilic $:D^-$ first attacks and reduces the carbonyl to form a basic alkoxide salt.

Problem 9.2 Prepare p–$ClC_6H_4CHOHCH_2CH_3$ from RCHO and R'MgX.

Since the groups on the carbinol C are different, there are two combinations possible.

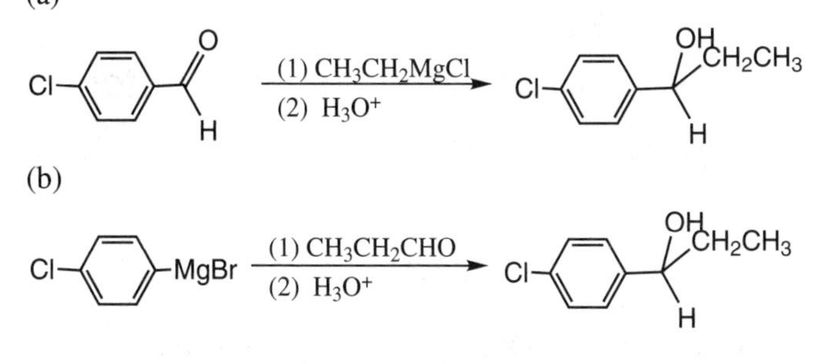

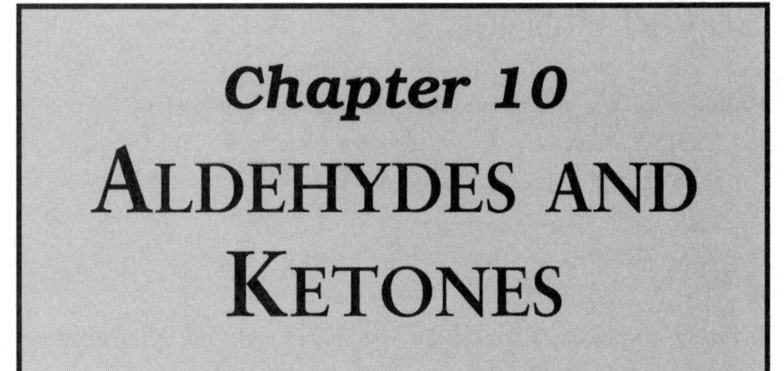

Chapter 10 ALDEHYDES AND KETONES

IN THIS CHAPTER:

Introduction and Nomenclature

Aldehydes and ketones have only H, R, or Ar groups attached to the **carbonyl group.** Aldehydes have at least one hydrogen bonded to the carbonyl group; ketones have only R's or Ar's.

Aldehyde names are based on the longest continuous chain including the carbon of the aldehyde group (the carbonyl carbon). The -e of the alkane name is replaced by the suffix -al. The carbon of –CHO is number 1. Common names for aldehydes replace the suffix -ic (-oic or -oxylic) and the word acid of the corresponding carboxylic acids by -aldehyde. The compound is named as an aldehyde whenever –CHO is attached to a ring.

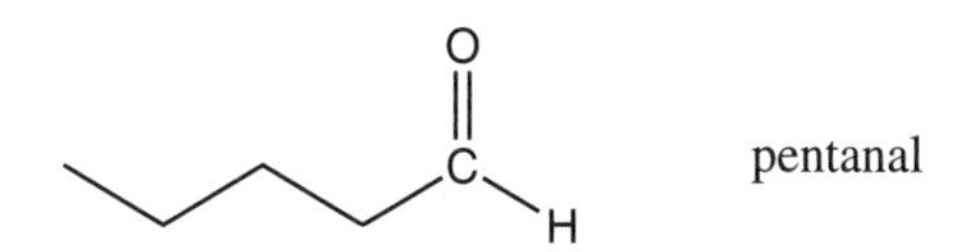

pentanal

Common names for ketones use the names of R or Ar as separate words, along with the word ketone. The IUPAC system replaces the -e of the name of the longest chain by the suffix –one, along with a number to indicate the position of the carbonyl group.

$CH_3COCH_2CH_3$ 2-butanone (or methyl ethyl ketone)

The electron-releasing alkyl groups (R) diminish the electrophilicity of the carbonyl C, lessening the chemical reactivity of ketones. Furthermore, the R's, especially large bulky ones, make approach of reactants to the C more difficult. For these reasons, ketones are less reactive than aldehydes.

Preparation of Aldehydes and Ketones

Ozonolysis of alkenes and cleavage of 1,2-diols afford carbonyl compounds. These reactions are discussed in Chapter 5. Ketones and aldehydes can also be produced by hydration of alkynes, reactions that are discussed in Chapter 5. **Friedel-Crafts acylations** of arenes with RCOCl in the presence of $AlCl_3$ give good yields of ketones. These reactions are discussed in Chapter 7.

Alcohols are the most important precursors in the synthesis of carbonyl compounds, being readily available. Ordinarily, H_2SO_4/CrO_3 (Jones' reagent) is used to oxidize 2° R_2CHOH to R_2CO. However, oxidizing 1° RCH_2OH to RCHO without allowing the overoxidation of RCHO to RCO_2H requires special reagents. The most common of these is pyridinium chlorochromate, PCC.

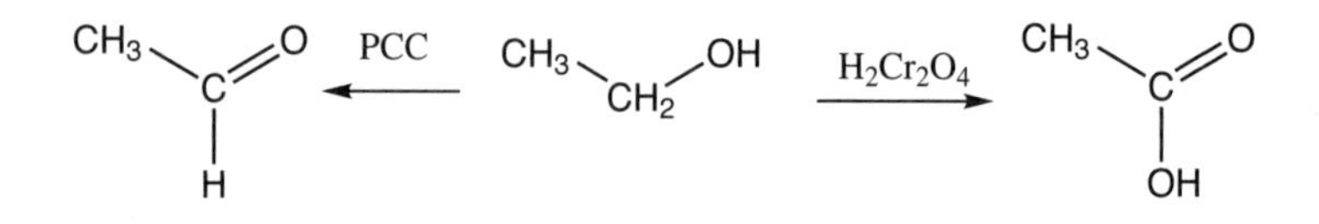

Acid chlorides, esters (RCO_2R'), and nitriles (RCN) are reduced with lithium tri-t-butoxyaluminum hydride, $LiAlH[OC(CH_3)_3]$ or with DIBAL (diisobutyl aluminum hydride, $AlH[CH_2CH(CH_3)_2]_2$), at very low temperatures, followed by H_2O workup. The net reaction is a displacement of X with H.

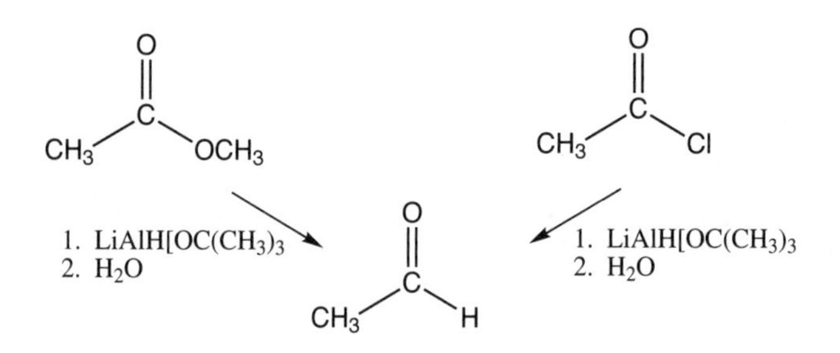

Reaction of Aldehydes and Ketones

Oxidation. Aldehydes undergo oxidation and ketones are unreactive under normal chromic acid oxidation conditions. A more mild oxidant is Tollens' reagent, $Ag(NH_3)_2^+$ (from Ag^+ and NH_3). Formation of the shiny Ag mirror in this test is a positive test for aldehydes. The RCHO must be soluble in aqueous alcohol. This mild oxidant permits –CHO to be oxidized in a molecule having groups more difficult to oxidize, such as 1° or 2° OH's.

Reduction. Aldehydes and ketones are readily reduced to 1° and 2° alcohols, respectively, with $NaBH_4$ or $LiAlH_4$.

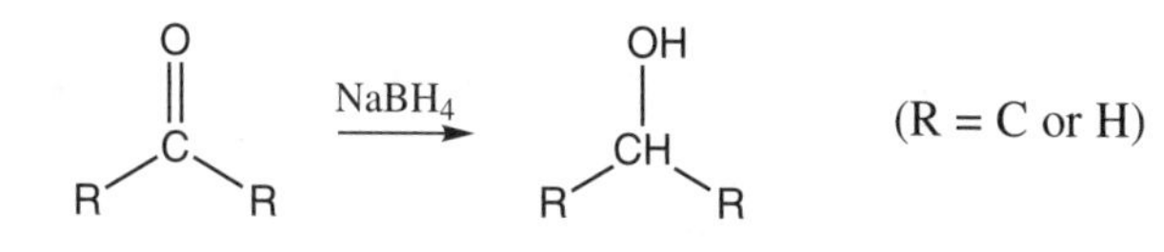

Addition Reactions of Nucleophiles to C = O

The carbon of the carbonyl group is electrophilic, due to a partial positive charge on the less electronegative carbon. This is easiest to see in the resonance hybrid of a carbonyl group. A variety of different nucleophiles will add to carbonyl groups, including **Grignard** reagents, which are discussed in Chapter 9.

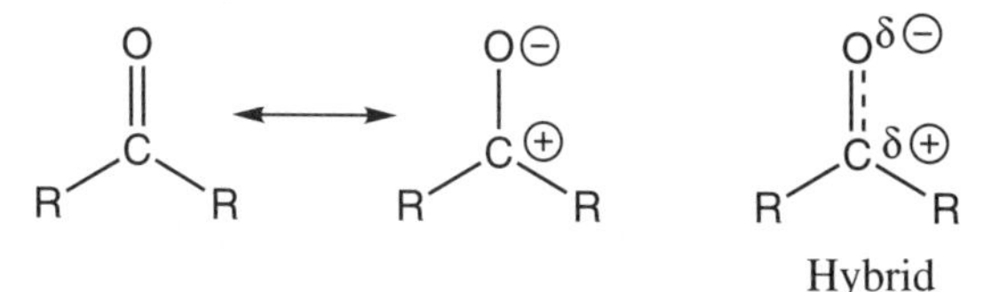

Acetal Formation. Alcohols add to aldehydes and ketones in an acid-catalyzed reaction to produce hemiacetals and acetals. In H_3O^+, RCHO is regenerated because acetals undergo acid-catalyzed cleavage much more easily than do ethers. Since acetals are stable in neutral or basic media, they are used to protect the –CH=O group.

> **Note!**
>
> **Enamines and Imine Formation**
>
> Amines are nucleophiles and add to aldehydes and ketones to form imines, which often isomerize to enamines. These reactions are discussed in Chapter 13.

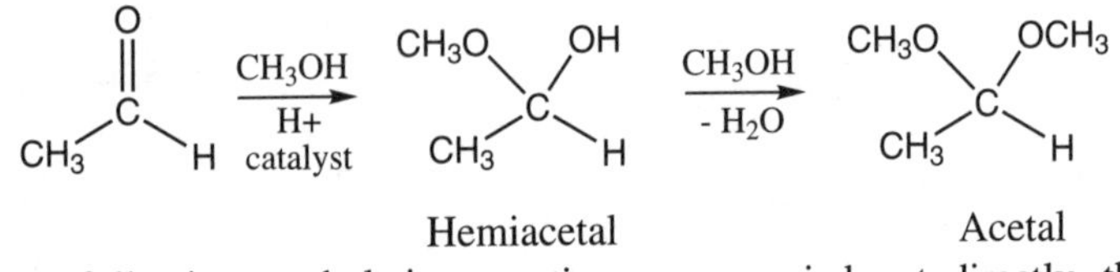

If the following methylation reaction were carried out directly, the acetylide ion would react as a nucleophile, attacking the carbonyl group. To prevent this side reaction, the carbonyl is protected by acetal formation before the carbanion is formed using the strongly basic sodium hydride (NaH). The acetal is stable under the basic conditions of the methylation reactions. The aldehyde is later unmasked by acid-catalyzed hydrolysis.

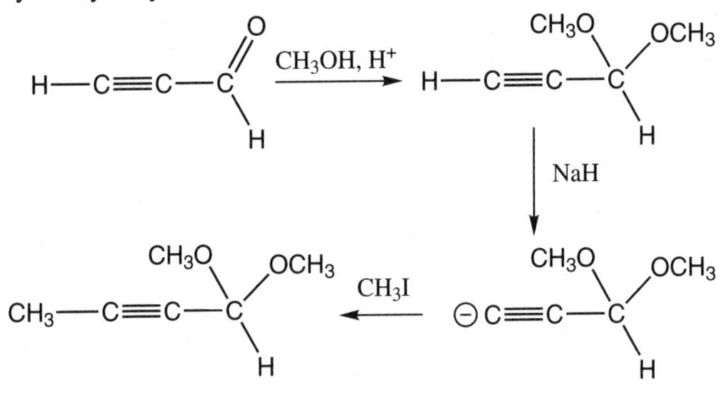

Ylides. A carbon anion can form a π bond with an adjacent phosphorus or sulfur. The resulting charge delocalization is especially effective if phosphorus or sulfur, furnishing the empty *d* orbital, also has a + charge. Carbon anions with these characteristics are called ylides.

The Wittig reaction uses phosphorus ylides to change the carbonyl group to an alkene. The ylide is prepared in two steps from RX.

$$Ph_3P\!:\; I\text{-}CH_2CH_3 \xrightarrow{-I^{\ominus}} Ph_3\overset{\oplus}{P}\text{-}CH_2CH_3 \xrightarrow{NaH} Ph_3\overset{\oplus}{P}\text{-}\overset{\ominus}{C}H_2CH_3$$

The ylide adds to the carbonyl group of the aldehyde (or ketone), and leads to an alkene and triphenylphosphine oxide. The cis-trans geometry of the alkene is influenced by the nature of the substituents, solvent, and dissolved salts. Polar protic or aprotic solvents favor the cis isomer.

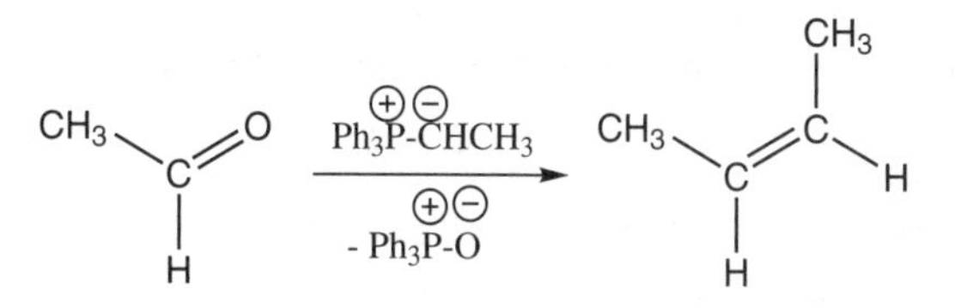

You Need to Know

- Preparation of ketones and aldehydes
- The electrophilic character of carbonyl groups
- Nucleophilic addition to carbonyls
- Reduction of ketones and aldehydes

Solved Problems

Problem 10.1 (a) What properties identify the carbonyl group of aldehydes and ketones? (b) How can aldehydes and ketones be distinguished?

(a) A carbonyl group (1) forms derivatives with substituted ammonia compounds such as NH_2OH, (2) shows strong IR absorption at 1690–1760 cm^{-1}.
(b) The C-H bond in aldehydes has a unique IR absorption at 2720 cm^{-1}. Aldehydes are oxidized with chromic acid, (turns reagent green), but ketones are not.

Problem 10.2 Show the product for the following reaction.
butanal + $HOCH_2CH_2OH$ (H^+ catalyst)

H O + HO OH $\xrightarrow{H^+}$ O O

Chapter 11
Carboxylic Acids and Their Derivatives

In This Chapter:

- ✔ *Nomenclature*
- ✔ *Carboxylic Acid Derivatives*
- ✔ *Preparation of Carboxylic Acids*
- ✔ *Acidity of Carboxylic Acids*
- ✔ *Reactions of Carboxylic Acids*
- ✔ *Polyfunctional Carboxylic Acids*
- ✔ *Reactions of Acid Derivatives*
- ✔ *Solved Problems*

Nomenclature

Carboxylic acids (RCO_2H or $ArCO_2H$) have the structure shown below. Some have names derived from acetic acid; e.g., $(CH_3)_3CCO_2H$ and $C_6H_5CH_2CO_2H$, are trimethylacetic acid and phenylacetic acid, respectively. Occasionally they are named as carboxylic acids, e.g., the compound here is cyclohexanecarboxylic acid.

For IUPAC names, replace the -e of the corresponding alkane with **-oic acid:** thus, $CH_3CH_2CO_2H$ is propanoic acid. The carbons are numbered; the carbon of CO_2H is numbered 1.

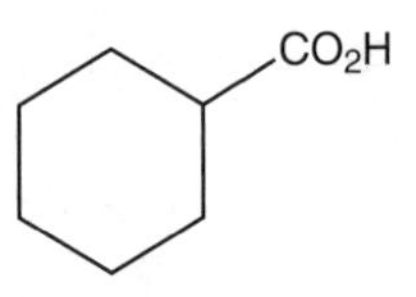

Carboxylic Acid Derivatives

The common types of acid derivatives are given in the table, with conventions of nomenclature that involve changes of the name of the corresponding carboxylic acid.

Preparation of Carboxylic Acids

Oxidation of 1° Alcohols, Aldehydes, and Arenes. Carboxylic acids can be prepared by oxidizing primary alcohols or aldehydes with CrO_3 in H_2SO_4 (**Jones oxidation**). Aromatic carboxylic acids can be made by side chain oxidation of substituted benzenes using $KMnO_4$.

Oxidative Cleavage of Alkenes and Alkynes.
Carboxylic acids can be prepared by oxidative cleavage of alkenes or alkynes using $KMnO_4$ in acid.

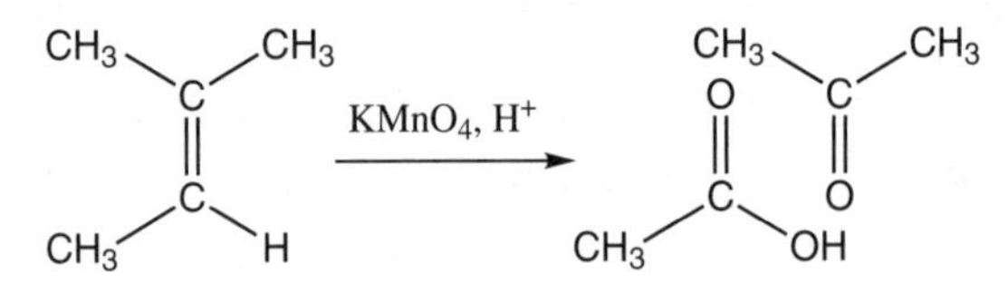

Grignard Reagent and CO_2. Addition of a Grignard reagent to CO_2 followed by acid workup leads to carboxylic acids.

$$RMgBr + CO_2 \longrightarrow RCO_2H$$

Hydrolysis of Acid Derivatives and Nitriles. Hydrolysis of carboxylic

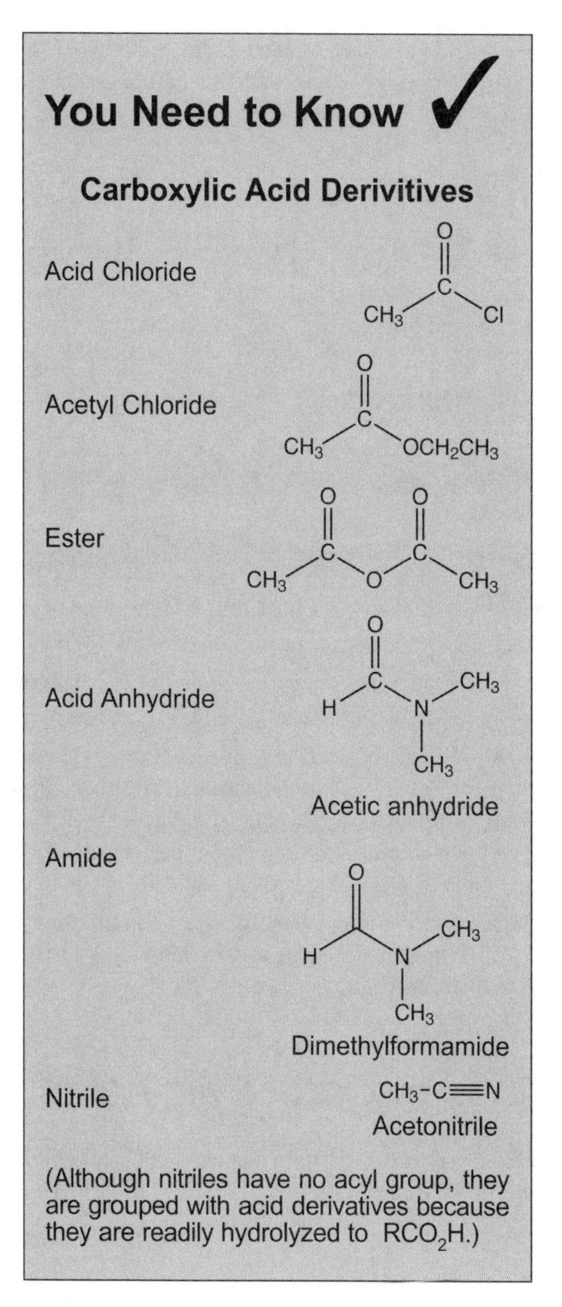
You Need to Know
Carboxylic Acid Derivitives
Acid Chloride
O
C
CH3
Cl
Acetyl Chloride
O
C
CH3
OCH2CH3
Ester
O
O
C
C
CH3
O
CH3
Acid Anhydride
O
C
H
N
CH3
CH3
Acetic anhydride
Amide
O
H
N
CH3
CH3
Dimethylformamide
Nitrile
CH3-C≡N
Acetonitrile
(Although nitriles have no acyl group, they are grouped with acid derivatives because they are readily hydrolyzed to RCO2H.)

acid derivatives (acid chlorides, esters, amides, anhydrides, and nitriles) using either acidic or basic water produces a carboxylic acid.

Acidity of Carboxylic Acids

The H of CO_2H is acidic because RCO_2^- is delocalized over both oxygens and is more stable and a weaker base than RO^-, whose charge is localized on only one oxygen.

$$RCO_2H + H_2O \longrightarrow RCO_2^{\ominus} + H_3O^{\oplus}$$

RCO_2H forms **carboxylate salts** with bases; when R is a long alkyl chain, these salts are called soaps.

$$RCO_2H + KOH \longrightarrow RCO_2^{\ominus}K^{\oplus} + H_2O$$

The influence of substituents on acidity is best understood in terms of the conjugate base, RCO_2^-, and can be summarized as follows. Electron-withdrawing groups stabilize the carboxylate anion, strengthening the acid. Electron-donating groups destabilize RCO_2^- and weaken the acid.

Like all halogens, Cl is electronegative, electron-withdrawing, and acid-strengthening. Since F is more electronegative than Cl, it is a better withdrawing group and a better acid strengthener.

Inductive effects diminish as the number of C's between Cl and the O's increases. $ClCH_2CO_2H$ is a stronger acid than $ClCH_2CH_2CO_2H$, since the chlorine is closer to the negative charge on the anion that it stabilizes. Two Cl's are more electron-withdrawing than one Cl, so Cl_2CHCO_2H is a stronger acid than $ClCH_2CO_2H$.

Reactions of Carboxylic Acids

Acid Chloride Formation. Carboxylic acids give acid chlorides when they are treated with thionyl chloride.

$$RCO_2H + SOCl_2 \longrightarrow RCOCl + HCl + SO_2$$

Reaction with $SOCl_2$ is particularly useful because the two gaseous products SO_2 and HCl are readily separated from RCOCl.

Ester Formation. Carboxylic acids react with alcohols to give esters. An acid catalyst is required.

$$RCO_2H + R'OH \xrightarrow{H^+} RCO_2R'$$

In this reaction, the oxygen of the C=O is protonated, which increases the electrophilicity of the carbonyl carbon and renders it more easily attacked in the slow step by the weakly nucleophilic R'OH. The tetrahedral intermediate undergoes a sequence of fast deprotonations and protonations, the end result being the loss of H^+ and H_2O and the formation of the ester.

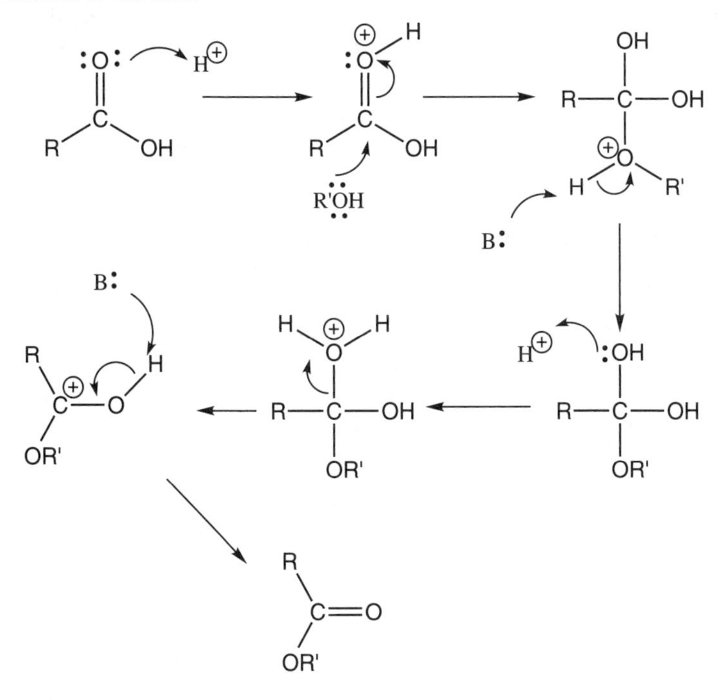

Reduction of Carboxylic Acids

Acids are best reduced to alcohols by $LiAlH_4$. $NaBH_4$ is not a strong enough reducing agent to reduce carboxylic acids.

Polyfunctional Carboxylic Acids

Dicarboxylic Acids. [$HO_2C(CH_2)n\ CO_2H$] The chemistry of dicarboxylic acids depends on the value of *n*. For $n = 1$, decarboxylations can occur upon heating the diacid. When $n = 2$ or 3, the diacid forms cyclic anhydrides when heated. Longer-chain α, ω-dicarboxylic acids usually undergo intermolecular dehydration on heating to form long-chain polymeric anhydrides.

Hydroxyacids: Lactones. Reactions of hydroxycarboxylic acids, $HO(CH_2)_nCO_2H$, also depend on value of *n*. In acid solutions, γ-hydroxycarboxylic acid ($n = 3$) and δ-hydroxycarboxylic acid ($n = 4$) form cyclic esters (lactones) with five-membered and six-membered rings, respectively. Intramolecular nucleophilic displacements, such as those in lactone formation, have faster reaction rates than intermolecular S_N2 reactions because the latter require two species to collide.

Reactions of Acid Derivatives

The more reactive derivatives are readily converted to the less reactive ones. Because acetic anhydride reacts less violently, it is often used instead of the more reactive acetyl chloride to make derivatives of acetic acid. Since other anhydrides are not readily available, the acid chlorides are used to make acid derivatives. The reactivity order is acid chlorides > anhydrides > ester > amide.

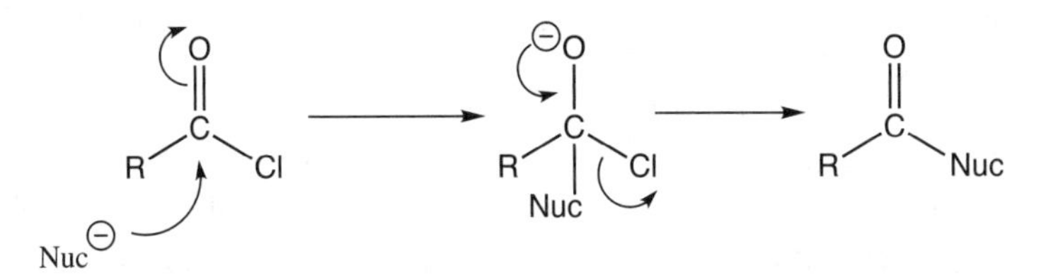

Nucleophilic substitution of acyl compounds takes place readily if the incoming group (Nu:⁻ or Nu:) is a stronger base than the leaving group (G:–) or if the final product is a resonance-stabilized RCO_2^-.

The reactions of acid derivatives generally involve nucleophilic attack at the carbonyl carbon. Nucleophilic substitutions of RCOG, such as RCOCl, occur in two steps. The first step (addition) resembles nucleophilic addition to ketones and aldehydes and the second step (elimination) is loss of G, in this case, Cl^-.

Reactions of this type can be carried out either in acid or in base. For example, in hydrolysis, the protonation of carbonyl O makes C more electrophilic and hence more reactive toward weakly nucleophilic H_2O. Strongly basic ^-OH readily attacks the carbonyl C. Unlike acid hydrolysis, this reaction is irreversible, because ^-OH removes H^+ from RCO_2H to form resonance-stabilized RCO_2^-.

Acid Chlorides. Acid chlorides are used in Friedel-Craft acylations of benzene rings with $AlCl_3$ catalyst, discussed in Chapter 7. They are also readily converted to other acid derivatives by reaction with the appropriate nucleophile.

Acid Anhydrides. Heating dicarboxylic acids, $HO_2(CH_2)_nCO_2H$ ($n = 2$ or 3), forms cyclic anhydrides by intramolecular dehydration. Intermolecular dehydration of carboxylic acids is used to prepare acetic anhydride, but other anhydrides are not readily formed using this method. Although they are less reactive than acid chlorides, anhydrides resemble acid halides in their reactions. Acid anhydrides can also be used to acylate aromatic rings in electrophilic substitutions.

Esters. Esters react more slowly than acid chlorides or anhydrides. They can be used to prepare amides by reaction with an amine.

Reduction of esters with $LiAlH_4$ gives alcohols, as discussed in Chapter 9.

Fats and Oils. Fats and oils are mixtures of esters of glycerol, $HOCH_2CHOHCH_2OH$, with acyl groups from carboxylic acids, usually with long carbon chains. These **triacylglycerols,** also called **triglycerides,** are types of **lipids** because they are naturally occurring and soluble only in nonpolar solvents. The acyl groups may be identical, or they may be different.

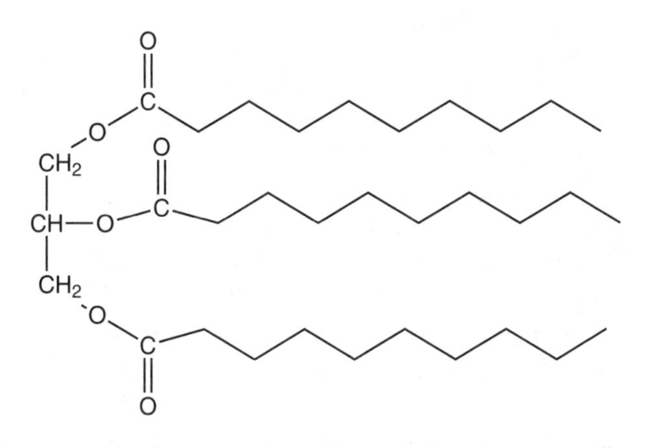

Amides. Unsubstituted amides may be prepared by careful partial hydrolysis of nitriles. Amides are slowly hydrolyzed under either acidic or basic conditions.

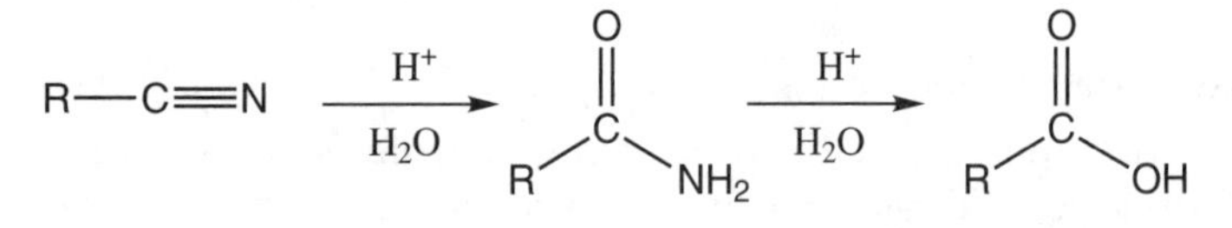

Imides. The hydrogen on N of the imides is acidic because the negative charge on N of the conjugate base is delocalized to each O of the two C=O groups, thereby stabilizing the anion.

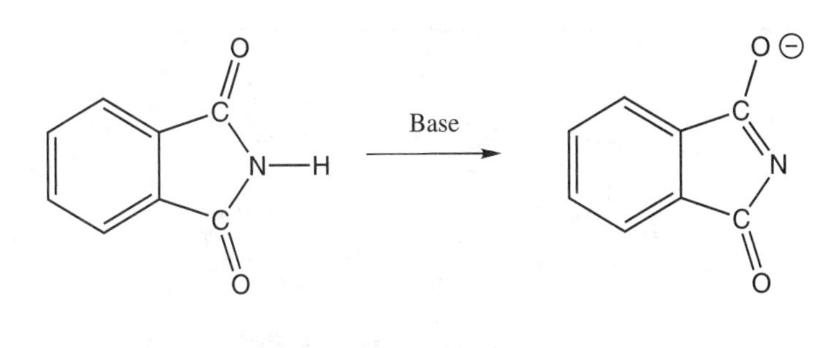

You Need to Know

- Preparation of carboxylic acids
- The resonance stabilization of the carboxylate anion
- Formation and reactions of acid derivatives

Solved Problems

Problem 11.1 Describe the electronic effect of C_6H_5 on acidity, if the acid strengths of $C_6H_5CO_2H$ and H CO_2H are 6.4 x 10^{-5} and 1.7 x 10^{-4}, respectively.

The weaker acidity of C_6H_5 CO_2H shows that the electron releasing resonance effect of C_6H_5 outweighs its electron attracting inductive effect.

Problem 11.2 Give the products from reaction of benzamide, $PhCONH_2$, with

(a) $LiAlH_4$, followed by H_3O^+

(b) hot aqueous NaOH

(a) $PhCH_2NH_2$

(b) $PhCO_2Na + NH_3$

Chapter 12
ENOLATES AND ENOLS

IN THIS CHAPTER:

Keto-Enol Tautomerization

Carbonyl compounds, ketones, and aldehydes in particular, are in rapid equilibrium with an isomer in which a hydrogen has moved from the α -carbon to the oxygen. This new isomer, which is both an alkene and an alcohol, is known as an enol. The keto form is usually the most stable.

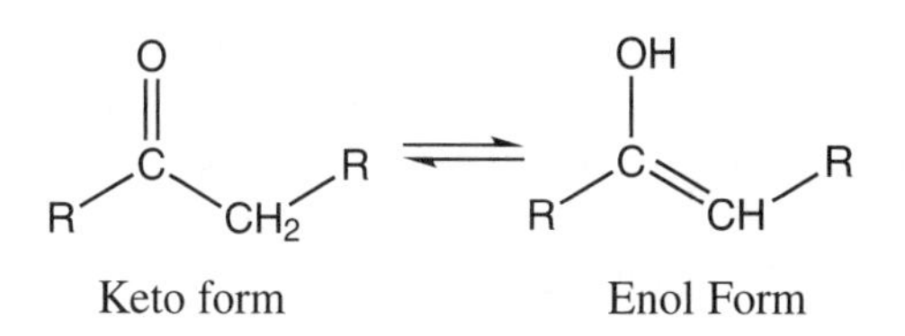

Structural isomers existing in rapid equilibrium are **tautomers** and the equilibrium reaction is **tautomerism.** The above is a keto-enol tautomerism.

Enolate Anions

Acidity of Hydrogens α to Carbonyl Groups. The carbon atoms immediately adjacent to a carbonyl group are called the α-carbons, and hydrogens on these carbons are acidic. The stability of the resulting anion is due to delocalization of the charge on to the oxygen atom. Acetone, for example, has a pK_a of 20.

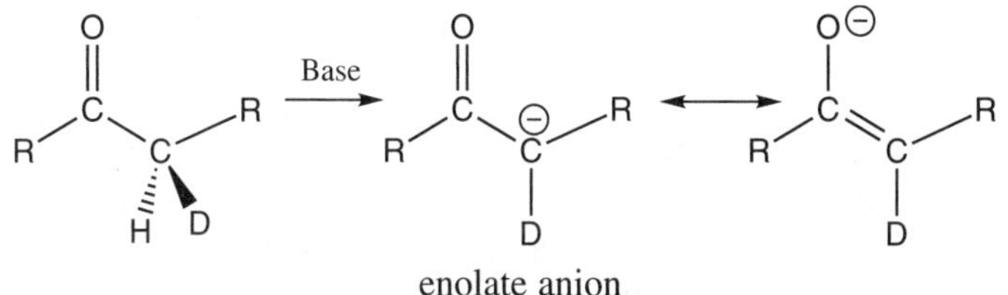

Racemization

The negatively charged carbon of the anion is no longer chiral, as it was in the reactant aldehyde, because the extended π bond requires a flattening of the conjugated portion of the anion of which this C is a part. Return of an H (or D) can occur equally well from the top face, to give one enantiomer, or from the bottom face, to give an equal amount of the other enantiomer; the product is racemic.

Treatment of ketones, aldehydes, esters, and amides (among other carbonyl-containing compounds) results in the formation of **enolate** anions. Since the carbanion-enolates are ambidentate (they have two different nucleophilic sites), they can be alkylated at carbon or at oxygen. Alkyl halides typically react preferably at carbon, while alkyl tosylates give larger amounts of O-alkylation.

Alkylation of Simple Enolates

Enolates are nucleophiles that react with alkyl halides (or sulfonates) by typical S_n2 reactions. Enolates are often formed using lithium diisopropylamide (LDA). This base is very strong and converts all the substrate to the anion. It is non-nucleophilic; it is too sterically hindered to react with RX.

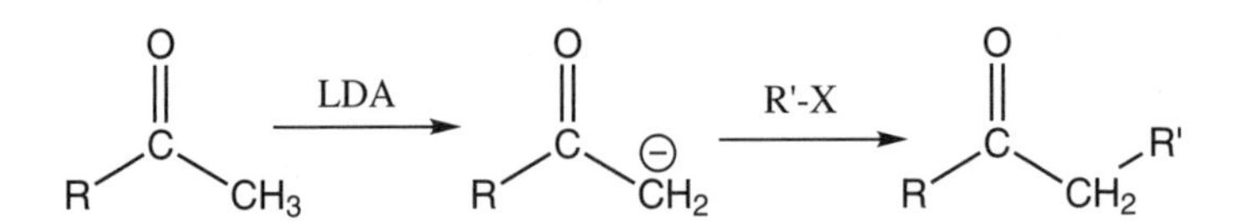

Undesired di- and tri-alkylation can occur if the anion is not produced quantitatively. Ketones with H's on more than one α–carbon can give a mixture of alkylation products. Several different approaches have been developed to circumvent these problems.

Enamine Alkylations. Monoalkylation is readily accomplished using this method. Enamines are made from a ketone and a secondary amine (R_2NH). Enamines of ketones are monoalkylated in good yield with reactive halides, such as benzyl and allyl. Enamines also can be acylated on the α–carbon with acid chlorides.

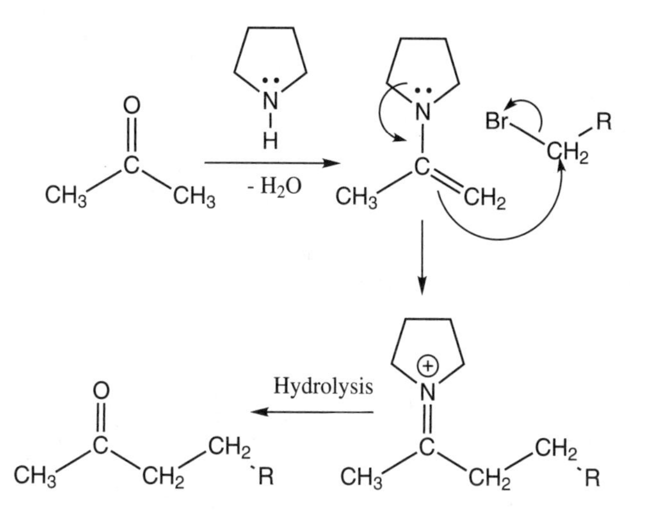

Alkylation Stable Enolates

The acidity of a hydrogen is greatly enhanced when the carbon to which it is attached is flanked by two carbonyl groups, as in diethyl malonate and ethyl acetoacetate. The anions formed from these compounds are stable and their reactivity is readily controlled.

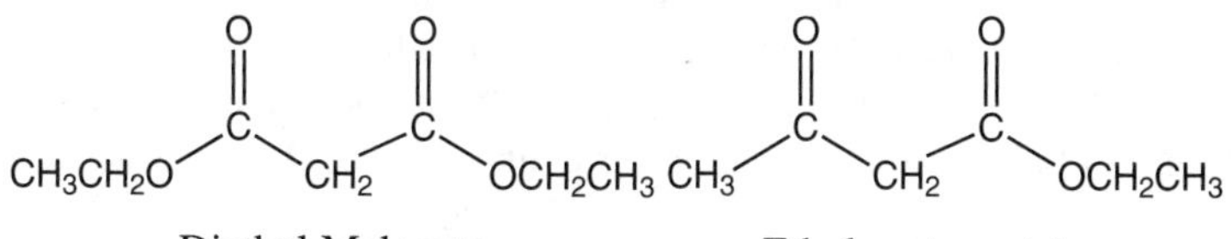

Diethyl Malonate Ethyl acetoacetate

Malonic Ester Synthesis of Substituted Acetic Acids. First, the enolate is formed with strong base (often NaOEt in EtOH), and the anion is alkylated by S_N2 reactions with unhindered RX or ROTs.

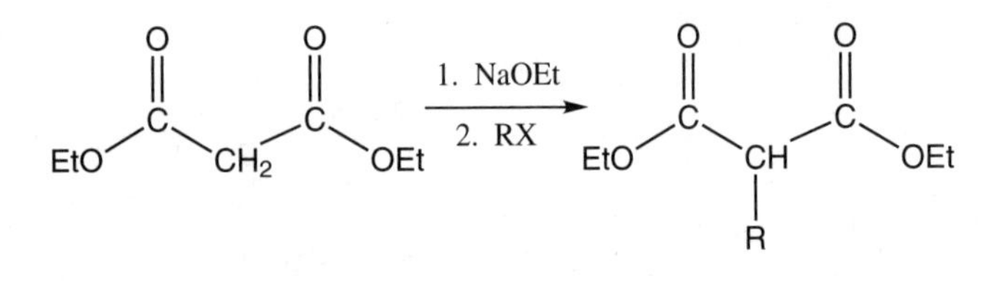

Hydrolysis of the substituted malonic ester gives the malonic acid, which undergoes **decarboxylation** (loss of CO_2) to form a substituted acetic acid.

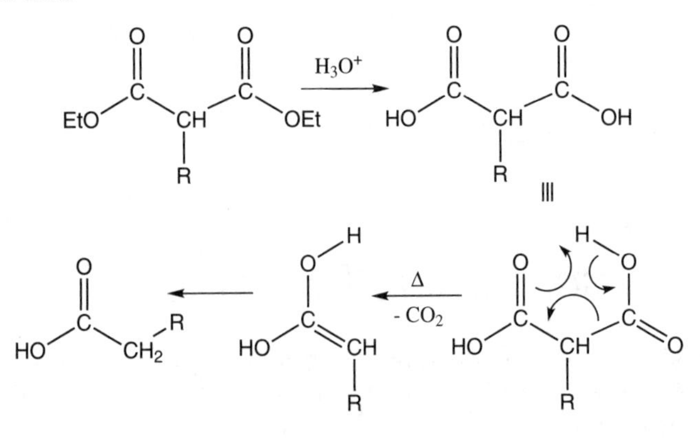

Acetoacetic Ester Synthesis. As with the malonic ester procedure, either one or two alkyl groups can be introduced in the acetoacetic procedure. The overall procedure is the same as in the malonic ester synthesis, and clean monoalkylation (or dialkylation, if 2 alkylation steps are used) results.

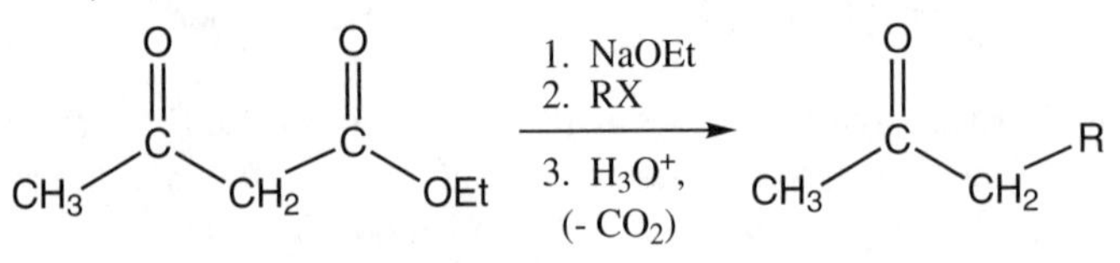

Nucleophilic Addition to Conjugated Carbonyl Compounds: The Michael Reaction

α, β –Unsaturated carbonyl compounds can be attacked by nucleophiles at the β carbon, forming an enolate.

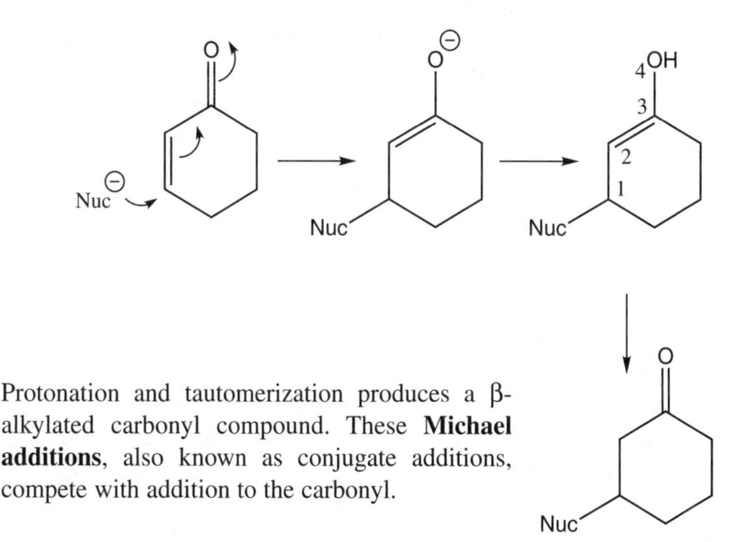

Protonation and tautomerization produces a β-alkylated carbonyl compound. These **Michael additions**, also known as conjugate additions, compete with addition to the carbonyl.

Condensations

A condensation reaction leads to a product with a new C–C bond. Most often the new bond results from a nucleophilic addition of a reasonably stable carbanion-enolate to the C=O group (acceptor) of an aldehyde; less frequently the C=O group belongs to a ketone or acid derivative. Another acceptor is the C≡N group of a nitrile.

Aldol Condensation. The addition of an enolate to the C=O group of its *parent compound* is called an **aldol condensation.** The product is a β-hydroxycarbonyl compound. In a **mixed aldol condensation**, the enolate of an aldehyde or ketone adds to the C=O group of a molecule other than its parent.

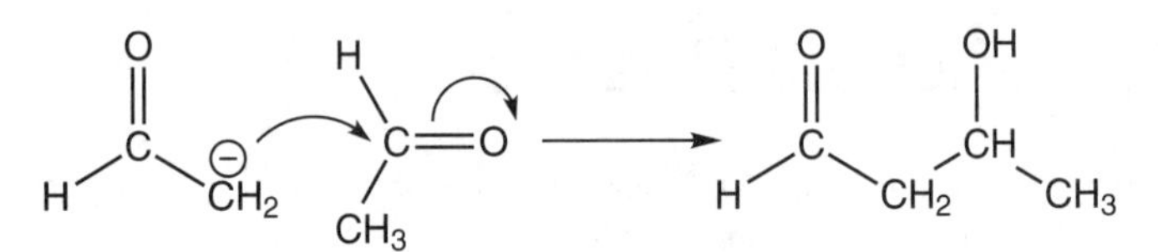

Aldol condensations are reversible, and with ketones the equilibrium is unfavorable for the condensation product. To carry out condensations of ketones, the product is continuously removed from the basic catalyst. β-Hydroxycarbonyl compounds are readily dehydrated upon heating to give α, β-unsaturated carbonyl compounds.

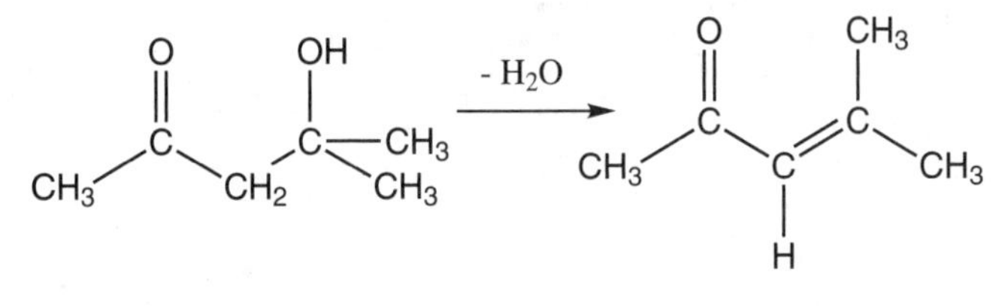

Aldol dimer of acetone

Acylation of Enolates: The Claisen condensation. In a **Claisen condensation**, the enolate of an ester adds to the C=O group of its parent ester. The addition is followed by loss of the OR group of the ester to give an α, β-ketoester. In a **mixed Claisen condensation**, the enolate adds to the C=O group of a molecule other than its parent. The sequence of reactions is similar to the reactions discussed above. An enolate is formed, which then adds to the carbonyl of an ester.

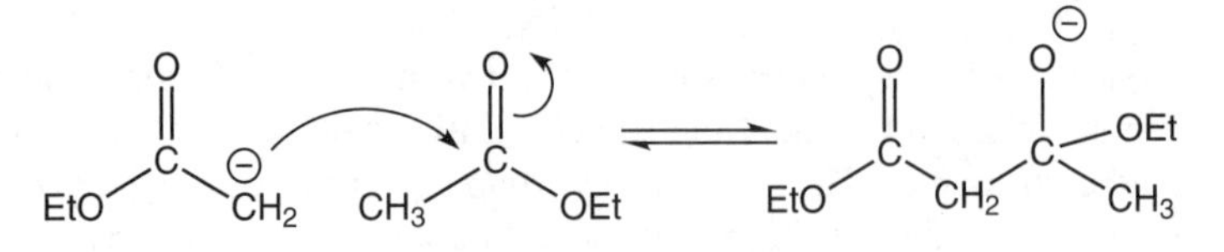

Next, the carbonyl group reforms, accompanied by the expulsion of the –OR portion of the ester.

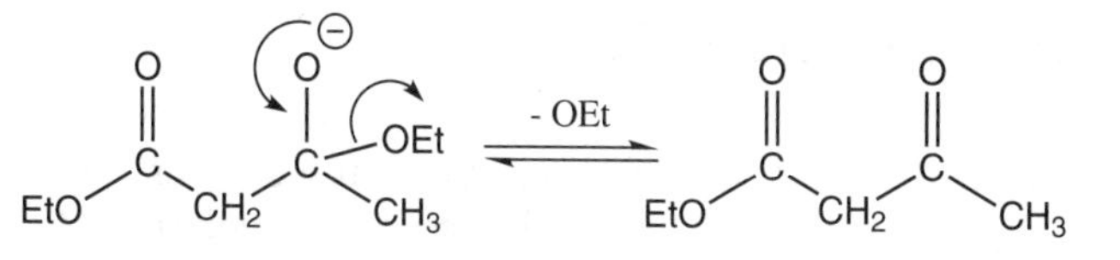

Each of these steps is reversible. The reaction is driven to completion by an irreversible step, in which a very stable enolate is formed. Acid is then added to protonate the enolate.

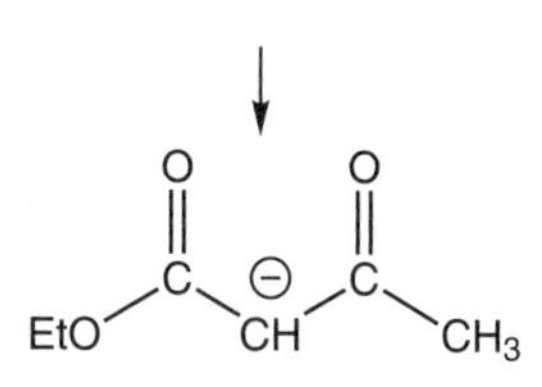

You Need to Know

- Keto-enol tautomerization
- Resonance stabilization of enolate anions
- Alkylation of enolates
- Malonic ester and acetacetic ester alkylations
- Claisen condensation and the Michael reaction

Solved Problem

Problem 12.1 The Robinsen "annulation" reaction for synthesizing fused rings uses Michael addition followed by intramolecular aldol condensation. Illustrate with cyclohexanone and methyl vinyl ketone, $CH_2{=}CHCOCH_3$.

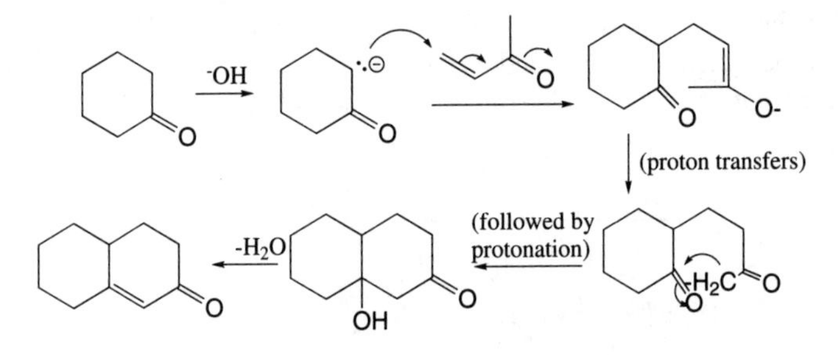

Chapter 13
AMINES

IN THIS CHAPTER:

- ✔ *Nomenclature*
- ✔ *Preparation*
- ✔ *Chemical Properties*
- ✔ *Reactions of Quaternary Ammonium Salts*
- ✔ *Diazonium Ions*
- ✔ *Spectral Properties*
- ✔ *Solved Problem*

Nomenclature

Amines are alkyl or aryl derivatives of NH_3. Replacing one, two, or three H's of NH_3 with alkyl groups gives **primary** (1°), **secondary** (2°), or **tertiary** (3°) amines, respectively. Examples are shown below. All four H's of NH_4^+ can be replaced with alkyl groups to give **quaternary** (4°) ammonium salts. Amines are named by adding the suffix -amine to the name of the alkyl group attached to N or to the longest alkane chain attached to N.

CH_3NH_2	$(CH_3)_2NH$	$(CH_3)_3N$
Methylamine a 1° amine	Dimethylamine a 2° amine	Triethylamine a 3° amine

Aromatic and cyclic amines often have common names such as aniline, pyridine, and piperidine.

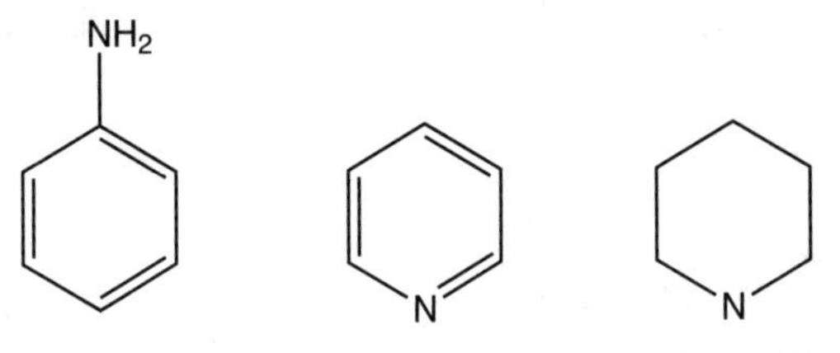

Aniline Pyridine Piperidine

Preparation

Alkylation of NH_3, RNH_2, and R_2NH with RX. Amines are nucleophiles, and hence they can be alkylated with alkyl halides. This reaction tends to produce a mixture of products, as each of the products are also reactive.

$$\ddot{N}H_3 \xrightarrow{CH_3I} CH_3\ddot{N}H_2 \xrightarrow{CH_3I} (CH_3)_2\ddot{N}H \xrightarrow{CH_3I} (CH_3)_3\ddot{N} \xrightarrow{CH_3I} (CH_3)_4\overset{\oplus}{N}\ \overset{\ominus}{I}$$

Reduction of N-Containing Compounds. Nitro compounds can be reduced with $LiAlH_4$, Zn/HCl, or H_2/Pt to produce the corresponding amine.

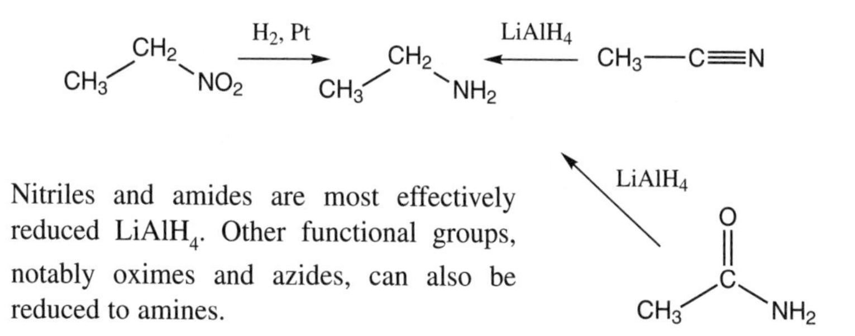

Nitriles and amides are most effectively reduced $LiAlH_4$. Other functional groups, notably oximes and azides, can also be reduced to amines.

Reductive Amination of Carbonyl Compounds. Ketones and aldehydes can be converted to amines in a 2 step procedure that involves an imine (sometimes called a **Schiff base**) intermediate. Imines are formed by addition of a 1° or 2° amine to the carbonyl, usually with removal of water to drive the reaction to completion. The imine can then be reduced, usually with sodium cyanoborohydride ($NaCNBH_3$) to produce the new amine.

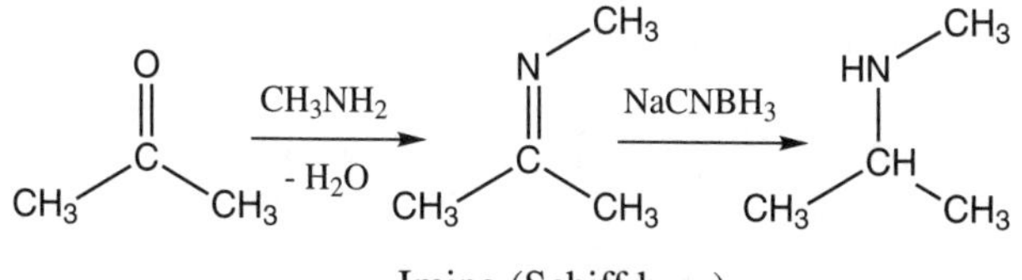

Imine (Schiff base)

Alkylation of Imides; Gabriel Synthesis of 1° Amines. To circumvent the difficulty of monalkylation of amines, the **Gabriel synthesis** is often used. Phthalimide is deprotonated with base and alkylated with R-X, then the imide is hydrolyzed to produce the monoalkyl amine.

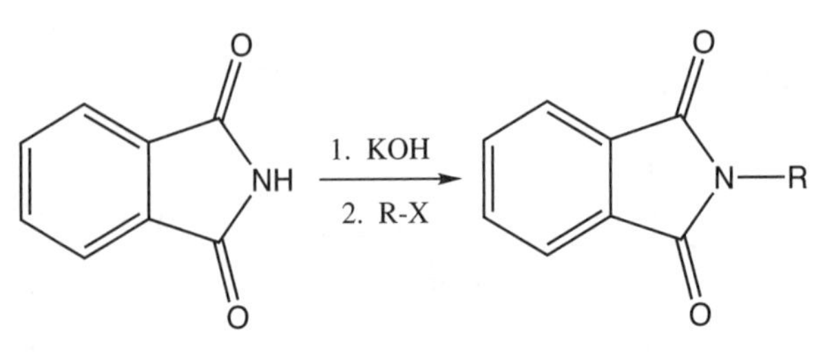

Phthalimide

KOH, H_2O
NH_2NH_2

NH_2—R

This method, even though it requires 2 steps, results in good yields of monoalkylamine.

Hofmann Degradation of Amides. Amides can be degraded by one carbon to produce amines by the Hofmann degradation. This procedure involves the treatment of the amide with Br_2, which results in an N-bromoamide. Deprotonation at nitrogen triggers a rearrangement in which a highly reactive, electron-deficient nitrene is formed, The alkyl R group migrates to nitrogen, resulting in an isocyanate. Hydrolysis and decarboxylation of the isocyanate produces the amine.

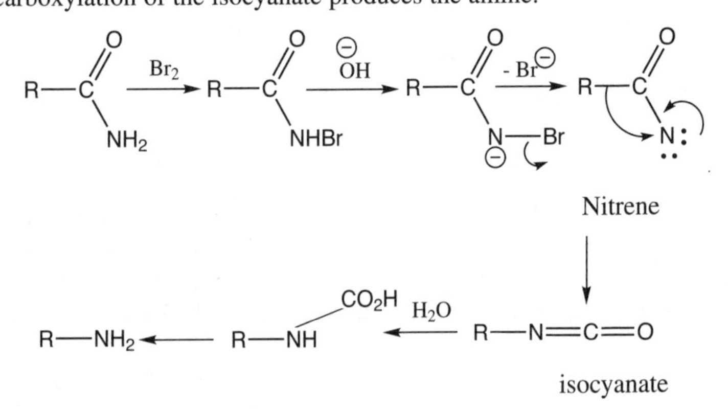

Chemical Properties

Stereochemistry. Amines with three different substituents and an unshared pair of electrons are chiral. However, in most cases, these chiral amines cannot be resolved. The amine undergoes a very rapid **nitrogen inversion** similar to that for a C undergoing an S_n2 reaction. This inversion of stereochemistry, essentially the rapid racemization of a chiral amine, is shown below.

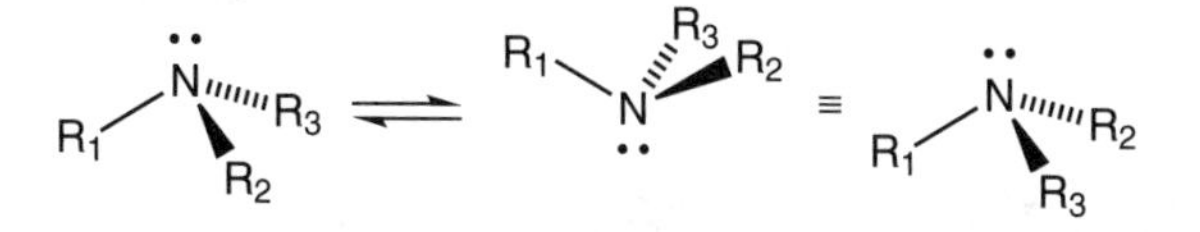

Acid-Base Chemistry. The lone pair of electrons on amines make them both nucleophiles and bases. Thus, amines will react with acids or with electrophiles.

$$Et_3N: + HCl \longrightarrow Et_3\overset{\oplus}{N}H\ \overset{\ominus}{Cl}$$

Acylation. 1° and 2° amines react with acid halides or esters to form amides. Overall this reaction results in transfer of the acyl group to the amine.

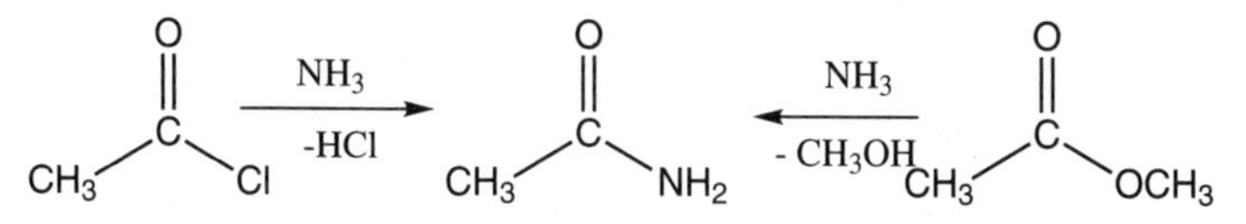

Reaction of 1° and 2° amines with benzenesulfonyl chloride (the **Hinsberg Reaction**) is quite similar to the reaction of these amines with carboxylic acid chlorides. The resulting sulfonamides are soluble in base if the amine was ammonia or a 1° amine [the remaining proton(s) are acidic]; but they are not soluble in base if the amine was 2°. Tertiary amines do not form stable products with benzenesulfonyl chloride.

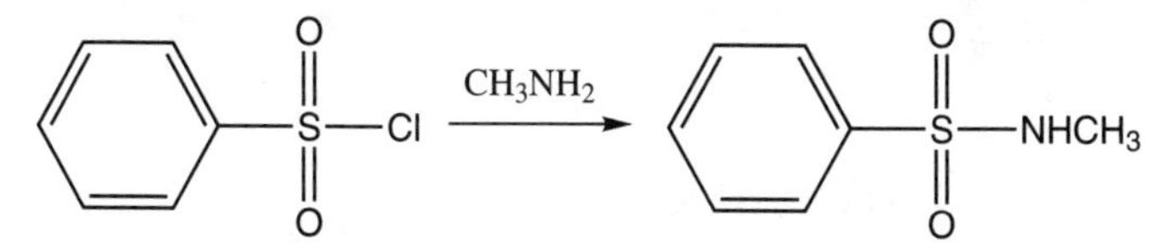

Reaction with Isocyanates. Isocyanates are reactive electrophiles and are readily attacked by NH_3 and 1° and 2° amines. The product of this addition is a urea. Again, 3° amines do not form stable products.

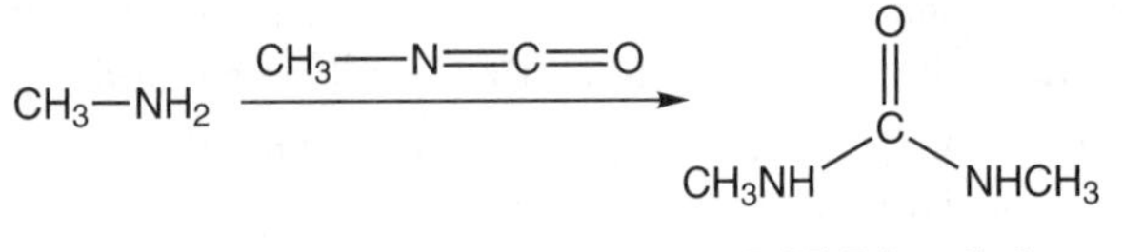

N, N'-Dimethylurea

Reactions of Quaternary Ammonium Salts

Quaternary ammonium salts are not nucleophilic or basic, since there is no lone pair of electrons on nitrogen as there is in amines. Upon heating in the presence of strong base, quaternary ammonium salts undergo the **Hofmann elimination**, an E2 elimination in which an alkene and a 3° amine are formed.

$$(CH_3CH_2)_4N^{\oplus} \xrightarrow{\text{NaOH}} (CH_3CH_2)_3N\text{:} + CH_2{=}CH_2$$

> **Note!**
>
> This E2 elimination gives the less substituted alkene (Hofmann product) rather than the more substituted alkene (Saytzeff product).

Diazonium Ions

Amines react with nitrous acid (HONO) to form diazonium ions. Alkyl diazonium ions decompose too quickly to be useful, but aryl diazonium ions have rich and highly useful chemistry. The N_2 molecule is an extremely good leaving group, so it can be substituted by a number of different nucleophiles. Aryl halides can be prepared by adding CuCl or CuBr to the diazonium ion (the **Sandmeyer** reaction). The N_2 can be replaced with a hydrogen by treatment of a diazonium ion with hypophosphorus acid (H_3PO_2). Examples of these substitution reactions of aryl diazonium ions are shown below.

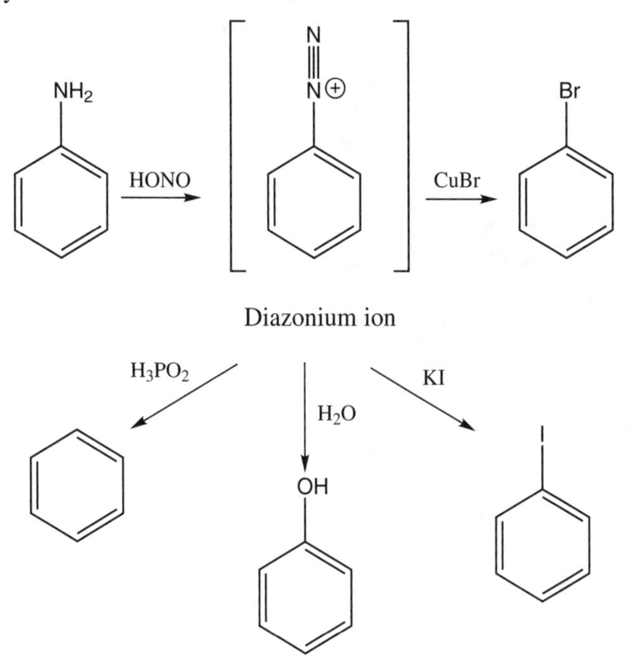

Spectral Properties

The N–H stretching and NH bending frequencies occur in the IR spectrum at 3050–3550 cm^{-1} and 1600–1640 cm^{-1} respectively. In the N–H stretching region, 1° amines and unsubstituted amides show a pair of peaks for a symmetric and an antisymmetric vibration. In the ^1H NMR spectrum, N–H proton signals of amines fall in a wide range (1 – 5 ppm) and are often very broad to due chemical exchange. The signals of N–H protons of amides, appearing at 5 –8 ppm, are even broader. The mass spectra of amines show an *odd* mass for parent ions if there is an odd number of nitrogens, and an *even* mass if there is an even number of nitrogens. For example, Et_3N has a molecular weight of 101.

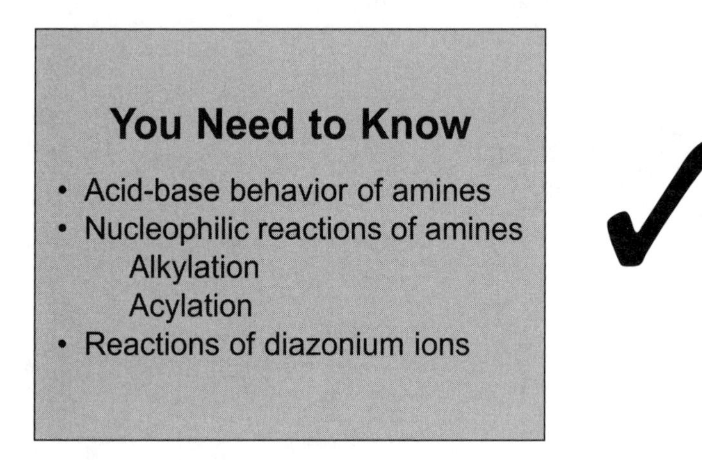

Solved Problem

Problem 13.1 Give the product of reaction in each case:
(a) C_2H_5Br + excess NH_3
(b) CH_2=CHCN + H_2/Pt

(a) $CH_3CH_2NH_2$
(b) $CH_3CH_2CH_2NH_2$

Chapter 14
Amino Acids, Peptides, and Proteins

In This Chapter:

- ✔ Introduction
- ✔ Chemical Properties of Amino Acids
- ✔ Peptides
- ✔ Structure Determination
- ✔ Peptide Synthesis
- ✔ Proteins
- ✔ Protein Structure
- ✔ Solved Problem

Introduction

This chapter deals with the very important α-**amino acids,** the building blocks of the **proteins** that are necessary for the function and structure of living cells. Enzymes, the highly specific biochemical catalysts, are proteins.

Amino acids are zwitterions: they have both + and - charges in them at neutral pH. The natural amino acids have different side-chains (R groups), but the same α-aminocarboxylic acid backbone. With the exception of glycine (R=H), the natural amino acids are chiral, with the S configuration at the α-carbon.

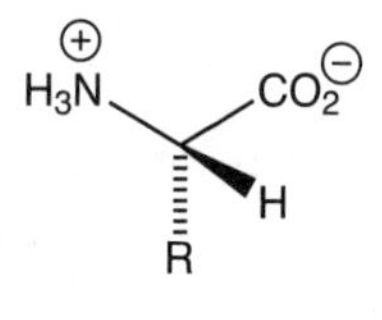

The 20 most common naturally occurring amino acids are listed below. They all are primary amines, except for proline which is a 2° amine. They are grouped into different catagories, depending on the chemical nature of the side-chain group. Each amino acid has a 3-letter abbreviation.

Natural α-Amino Acids

Type	Symbol	R
Neutral		
Glycine	Gly	–H
Alanine	Ala	$–CH_3$
Valine	Val	$–CH(CH_3)_2$
Leucine	Leu	$–CH_2CH(CH_3)_2$
Proline	Pro	$–CH_2CH_2CH_2$-
Isoleucine	Ileu	$-CH(CH_3)CH_2CH_3$
Serine	Ser	$-CH_2OH$
Threonine	Thr	$-CH(OH)CH_3$
Acidic and Amide		
Aspartic acid	Asp	$–CH_2CH_2CO_2H$
Asparagine	Asn	$–CH_2CH_2CONH_2$
Glutamic acid	Glu	$–CH_2CH_2CH_2CO_2H$
Glutamine	Gln	$–CH_2CH_2CH_2C(O)NH_2$

Basic		
Lysine	Lys	$-CH_2CH_2CH_2CH_2NH_2$
Arginine	Arg	$-CH_2CH_2CH_2NH-C(NH)NH_2$
Histidine	His	$-CH_2-$ (imidazol-4-yl: ring with H–N and N)
Sulfur-Containing		
Cysteine	Cys	$-CH_2SH$
Methionine	Met	$-CH_2CH_2SCH_3$
Aromatic		
Phenylalanine	Phe	$-CH_2C_6H_5$
Tyrosine	Tyr	$-CH_2C_6H_4OH$
Tryptophan	Trp	$-CH_2-$ (indol-3-yl, N–H)

Chemical Properties of Amino Acids

Acid-Base (Amphoteric) Properties. The pH at which [anion] = [cation] is called the **isoelectric point**. At this pH, there is no net charge on the molecule. The different amino acids have different isoelectric points, depending on whether the side-chain is acid, basic, or neutral.

Peptides

Peptides are polyamides composed of the different amino acids. The amide bond between two **amino acid residues** is often called a **peptide**

bond. Peptides can range from simple 2-amino acid residue compounds up to long chains of 50 to 100 residues, at which point they approach the size of proteins.

The artificial sweetener **aspartame** is a synthetic dipeptide, the methyl ester of aspartylphenylalanine, or AspPheOCH_3.

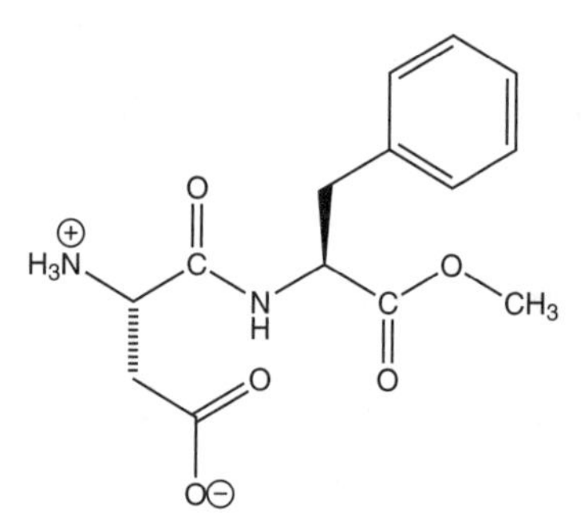

Structure Determination

It is possible to completely hydrolyze a peptide to its component amino acids by heating in aqueous acid. While this can, in principal, provide the composition of the peptide, it does not reveal the order in which the amino acids occur in the chain. This information (the sequence) is determined by other methods.

Sequencing. By cutting off one terminal amino acid at a time from either the free amino end (N-terminus) or the free COOH end (C-terminus), sequencing techniques give the linkage order (primary structure) of the amino acids in the peptide.

The N-terminal residue can be identified with either the **Sanger** method or the **Edman** method. In the Sanger method, the free NH_2 group of the N-terminal residue reacts with 2,4-dinitrofluorobenzene (DNFB) in an unusual example of *nucleophilic* aromatic substitution. After complete hydrolysis of the peptide, the DNFB-labelled amino acid can be isolated and identified.

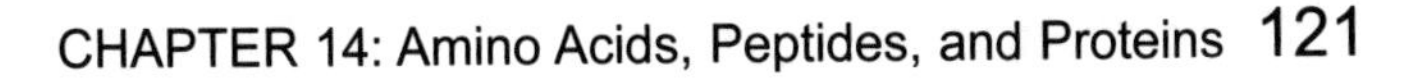

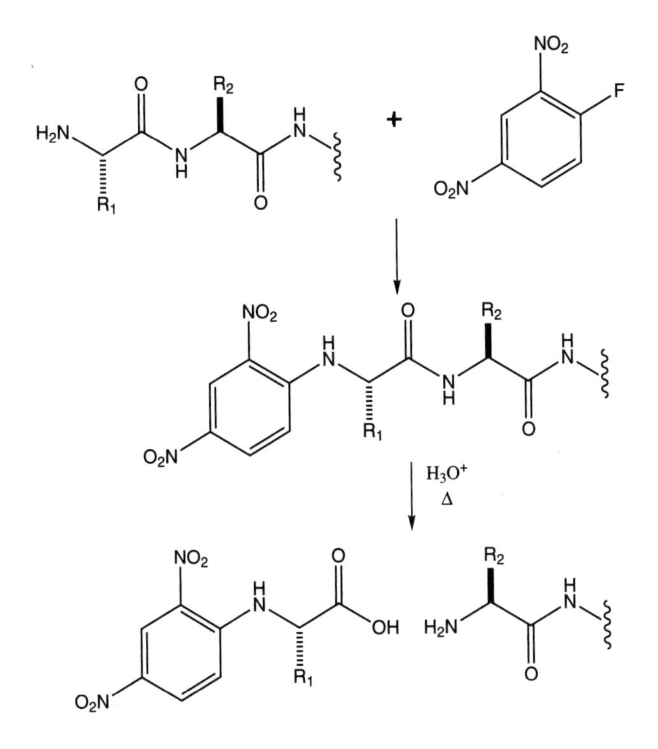

The Edman degradation is similar, although phenylisothiocyanate is used to label the N-terminal residue, and a phenylthiohydantoin is isolated. The Edman degradation can be repeated to identify each residue in turn.

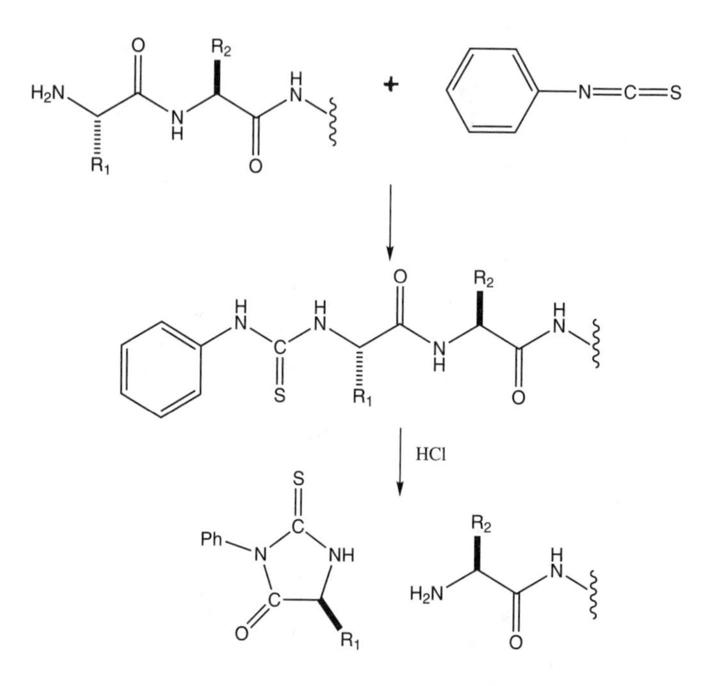

Carboxypeptidase, an enzyme, selectively cleaves the C-terminal amino acid from a peptide chain, and so can be used to identify each residue in turn.

Peptide Synthesis

The synthesis of peptides from amino acids requires the use of **protecting groups** to prevent random coupling of amino acids and polymerization. These groups can be installed to render a functional group unreactive, then removed to liberate the reactive functional group.

Blocking the N-Terminal Amino Group. This reaction is achieved by attaching a group (B) to N of NH_2. After the desired peptide is constructed, the blocking group is removed without destroying the peptide linkages. A common group is the t-butoxycarbonyl, or "Boc" group:

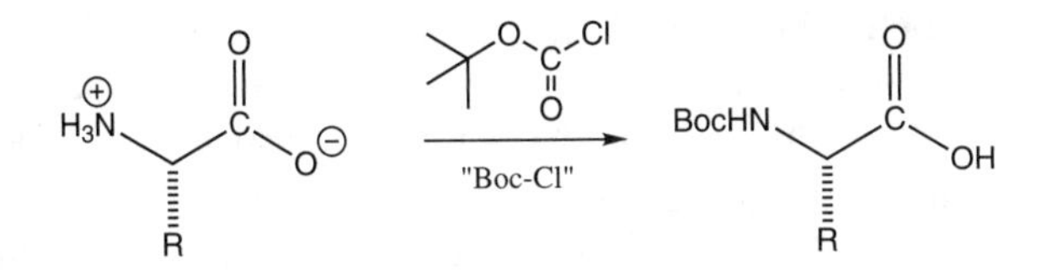

Activating the COOH. The –COOH is not reactive enough on its own to form an amide. It is first activated to form a peptide bond with an N-terminal NH_2 by reacting the N-blocked amino acid with ethyl chloroformate (EtOC(O)Cl). A mixed anhydride of an alkyl carbonic acid is formed. This technique minimizes racemization of the chiral α-carbon

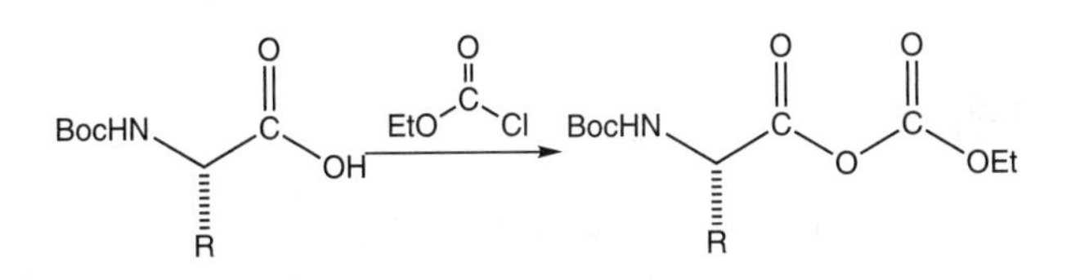

The COOH groups can also be activated in the presence of free N-terminal amino groups with N,N-dicyclohexylcarbodimide (DCC).

Coupling. A CO_2H-activated, Boc protected amino acid can be coupled with an amino acid with a free NH_2 group to make a dipeptide.

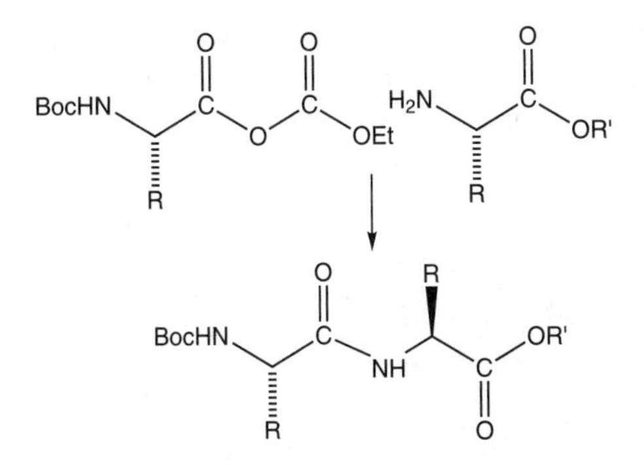

The synthesis can be continued, or the protecting groups can be hydrolyzed off with aqueous acid to liberate the free peptide.

Merrifield Solid-Phase Synthesis. A major advance in peptide synthesis is the solid-phase method. Beads of solid chloromethylated polystyrene [chloromethyl groups ($–CH_2Cl$) para to about every hundredth phenyl group] are used in an automated process. The sequence of steps is: (1) a C-terminal Boc-protected amino acid is attached by substitution for the Cl of the $–CH_2Cl$ group; (2) the Boc group is removed; (3) a Boc-protected amino acid is added and the new peptide bond is formed with the aid of DCC. Steps (2) and (3) are repeated as many times as is necessary to make the desired peptide. In the final step, the completed peptide is detached by hydrolysis of the ester bond that holds the peptide to the solid surface. The average yield for each step exceeds 99%.

Proteins

Proteins are polypeptides with molecular weights from ca. 10,000 up to several million, and are a major constituent of living cells. Simple proteins are hydrolyzed to amino acids. **Conjugated proteins** are composed of amino acids and nonpeptide substances known as **prosthetic groups**. These prosthetic groups include nucleic acids of **nucleoproteins**, carbohydrates of **glycoproteins**, pigments (such as **hemin** and **chlorophyll**) of **chromoproteins,** and fats or lipids of **lipoproteins.**

Amphoteric Properties. Isoelectric Points and Electrophoresis. Proteins have different isoelectric points, and in an electrochemical cell they migrate to one of the electrodes (depending on their charge, size, and shape) at different speeds. This difference in behavior is used in electrophoresis for the separation and analysis of protein mixtures.

Remember!

Hydrolysis

Proteins are hydrolyzed to their α-amino acids by heating with aqueous strong acids or, at room temperature, by digestive enzymes such as trypsin, chymotrypsin, or pepsin.

Protein Structure

The **primary structure** of a protein consists of the sequence of the constituent amino acids. The **secondary structure** arises from different conformations of the protein chains; these conformations are best determined by X-ray analysis. There are three types.

(a) The α-helix is a mainly right-handed coiled arrangement maintained by H-bonds between an N–H and O=C that are four peptide bonds apart.

(b) The **pleated sheet** has chains lying side by side and linked through N–H – – – – O=C H bonds. The α C's rotate slightly out of the plane of the merger (to minimize repulsions between their bulky R groups), which gives rise to the "pleats."

(c) **Random** structures have no repeating geometric pattern. However, there are structural constraints on the randomness leading to a constrained random orientation.

Tertiary structure is determined by any folding of the chains. There are two types.

(a) **Fibrous** proteins are water-insoluble, elongated, threadlike helixes (occasionally pleated sheets) made up of chains which are bundled together *intermolecularly* through N–H – – – – O=C H-bonding. They include **fibroin** (found in silk), **keratin** (in hair, skin, feathers, etc.), and **myosin** (in muscle tissue).

(b) **Globular** proteins, or globulins, are folded into compact spheroid shapes such that hydrophilic R groups point outward toward the water solvent and the hydrophobic (lipophilic) R groups turn inward. As a result, globulins can dissolve in or easily emulsify with water. The shape is maintained by *intramolecular* H bonding. The secondary structure is a combination of random (always present), helical, and pleated structures. Globulins include all enzymes, antibodies, albumin of eggs, hemoglobin, and many hormones such an insulin.

Quaternary structure exists when two or more polypeptide chains are linked only by weak forces of attraction among R groups at the *surface* of the chains. Such proteins are called oligomers (dimers, trimers, and so on).

Denaturation. Heat, strong acids or bases, ethanol, or heavy-metal ions irreversibly alter the secondary structure of proteins (see below). This process, known as **denaturation,** is exemplified by the heat-induced coagulation and hardening of egg white (albumin). Denaturation destroys the physiological activity of proteins.

You Need to Know

- The structure of α-amino acids
- Peptide and protein structure
- Peptide synthesis

Solved Problem

Problem 14.1 Account for the stereochemical specificity of enzymes with chiral substrates.

Since enzymes are proteins made up of optically active amino acids, enzymes are themselves optically active and therefore react with only one enantiomer of a chiral substrate.

Chapter 15
CARBOHYDRATES AND NUCLEIC ACIDS

IN THIS CHAPTER:

Introduction

Carbohydrates (**saccharides**) are aliphatic polyhydroxyaldehydes (**aldoses**), polyhydroxyketones (**ketoses**), or compounds that can be hydrolyzed to them. The suffix -ose denotes this class of compounds. The **monosaccharide** D-(+)-glucose, an aldohexose, is formed by plants in photosynthesis and is converted to the **polysaccharides** cellulose and starch. Simple saccharides are called **sugars**, and typically have

molecular formulae of $C_n(H_2O)_n$. Polysaccharides are hydrolyzable to monosaccharides. A carbohydrate is classified as either a **ketose** (if it contains a ketone carbonyl group) or an **aldose** (if it contains an aldehyde carbonyl group).

Fisher Projections. A simplest carbohydrate is glyceraldehyde. It is an aldotriose, since it is a 3-carbon aldehyde. It is common to draw carbohydrates in a **Fisher projection**, in which the carbon chain is drawn vertically on the page (aldehyde at top), and each stereogenic center is arranged so that the carbon chain is oriented *back into the page.* The substituents (hydrogen and hydroxyl) are drawn horizontally. The example below is D-glyceraldehyde.

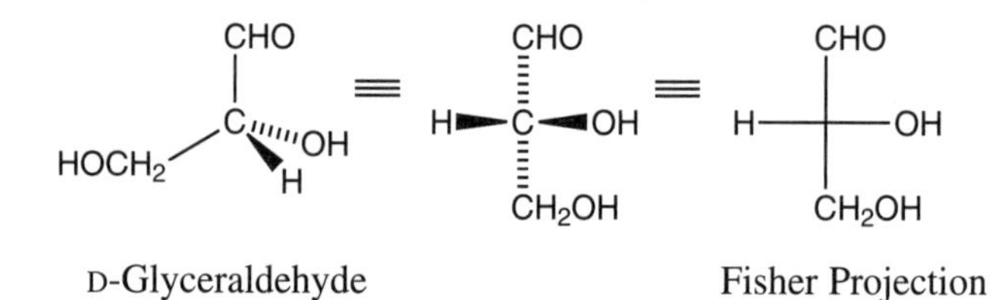

D-Glyceraldehyde Fisher Projection

Glucose is an aldohexose, so it is a 6-carbon aldehyde. The naturally occuring enantiomer, D-glucose, is shown here. The designation D stems from the last stereogenic center in the chain. If in the Fisher projection, the hydroxyl of the last stereogenic center (in the box) is on the right side of the chain, it is designated D. If that hydroxyl is on the left, the molecule is an L-sugar.

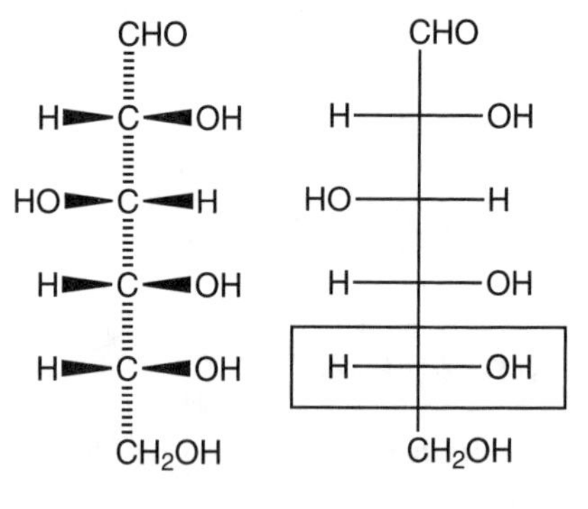

Mutarotation. Nearly all carbohydrates are chiral, so they rotate plane-polarized light either clockwise (+) or counterclockwise (–). Naturally occurring (+)-glucose is obtained in two forms: mp = 146 °C, $[\alpha]_D$ = +112° and mp = 150 °C, $[\alpha]_D$ = +19°. The specific rotation of each form of glucose changes (**mutarotates**) in water, and both reach a constant value of +52.7°. This phenomenon stems from the formation and equilibration of intramolecular hemiacetals. Carbohydrates that form 5-membered ring acetals are known as **furanoses** and ones that form 6-membered ring acetals are known as **pyranoses**.

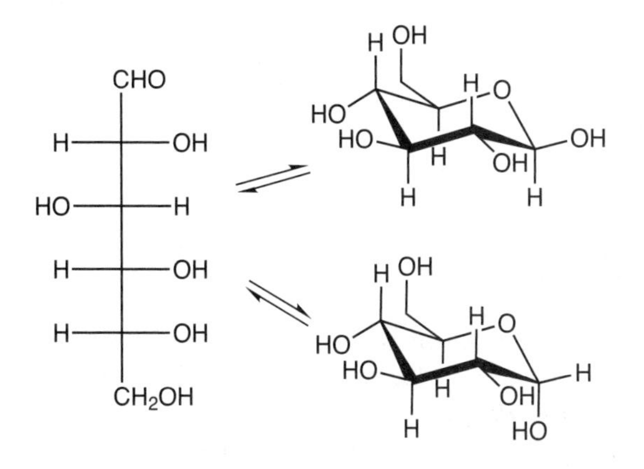

The two stereoisomeric hemiacetals are known as **anomers**, and they differ in configuration at the new stereogenic center formed from the carbonyl carbon. The anomer with the new hydroxyl "up" is known as the β-anomer, while, the anomer with the new hydroxyl "down" is the α-anomer.

Chemical Reactions of Monosaccharides

The chemistry of monosaccharides is the chemistry of the component aldehydes, ketones, and alcohols that make up these molecules. One example of how the chemistry of carbohydrates is the same as the chemistry of ketones is in their reaction with $NaBH_4$.

Reduction to Alditols. Both ketoses and aldoses can be reduced with $NaBH_4$ to reduced forms, known as **alditols**. With ketoses, this often produces 2 diastereomeric alditols.

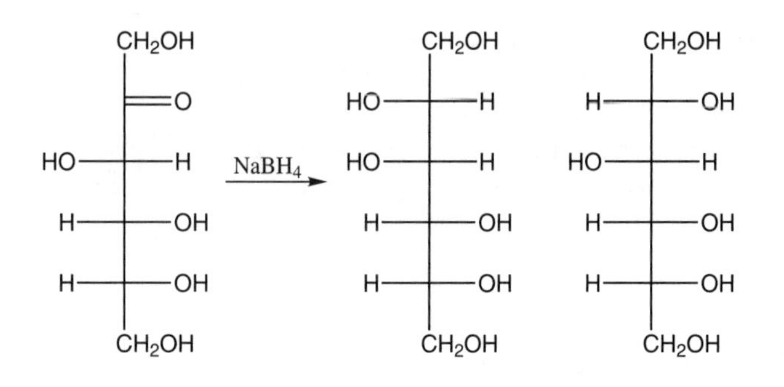

Disaccharides

Disaccharides are acetals in which an OH of one monosaccharide (the aglycone, denoted A) is bonded to the anomeric C of a second monosaccharide, B. The disaccharide is a **glycoside** of B, involving the anomeric center of one carbohydrate and an alcohol (in this case another carbohydrate). For example, the α-1,4-linked glucose dimer below is maltose. **Polysaccharides** are sugar polymers, linked in the same kind of manner.

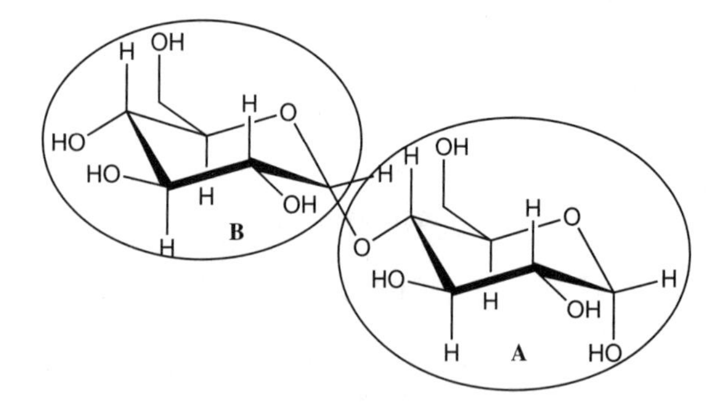

Nucleic Acid

The nucleic acids, **RNA** (ribonucleic acid) and **DNA** (deoxyribonucleic acid), are carbohydrate biopolymers with phosphate backbones. The repeating sugar in RNA is ribose, and in DNA it is 2-deoxyribose.

Nucleosides are glycosides of ribofuranose or deoxyribofuranose. Ribofuranose is the monosaccharide ribose in the furanose form. The nitrogenous bases (below) are bonded to the anomeric carbon (C_1) of the sugar.

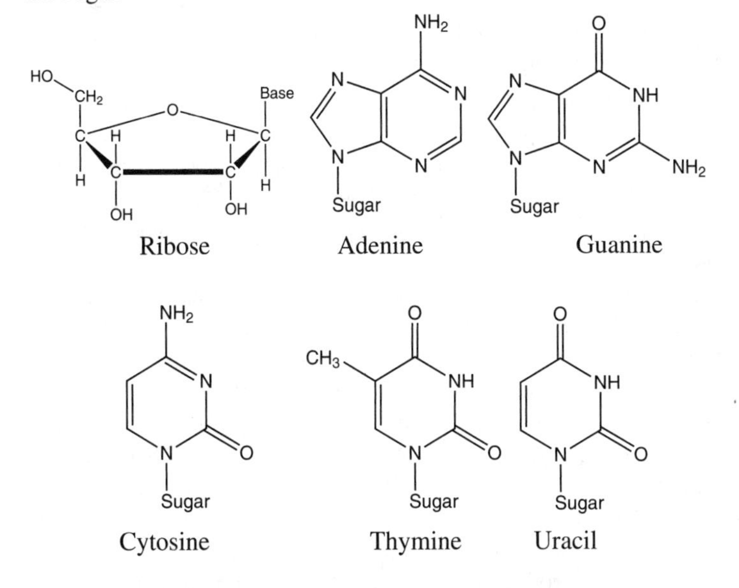

Adenine and guanine are known as the **purine** bases. Cytosine, thymine, and uracil are the **pyrimidine** bases.

Nucleotides are phosphate esters of nucleosides, formed at the CH_2OH group of the sugar. Adenosine triphosphate (**ATP**) is the triphosphate ester of adenosine, formed from adenine and ribose.

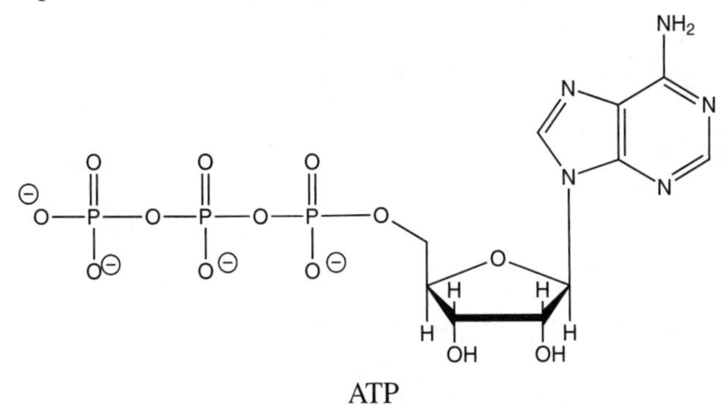

ATP

Nucleic acids are polymeric nucleotides in which phosphate esters link ribose or deoxyribose molecules through the C_1–OH of one and the C_3–OH of the other. In RNA, the aglycone nitrogen bases are cytosine, adenine, guanine, and uracil. In DNA, thymine replaces uracil. The RNA polymer is like that of DNA, except that in DNA an H replaces the OH group on C_2 of the ribose ring.

DNA, a constituent of the cell nucleus, consists of two strands of polynucleotides that are coiled to form a double helix. The strands are held together by H bonding between the nitrogen bases. The pyrimidines always form H bonds with a specific purine; i.e., cytosine with guanine and thymine with adenine. However, in RNA the pairing is between uracil and adenine.

Through its sequence of nitrogen bases, DNA stores the genetic information for cell function and division, and for biosynthesis of enzymes and other essential proteins. In protein synthesis, the information in the DNA is **transcribed** onto **messenger RNA** (mRNA), which moves from the nucleus to the ribosomes in the cytoplasm of the cell. Here the information is transferred to ribosomal RNA (rRNA). **Transfer RNA** (tRNA) carries amino acids to the surface of the rRNA, where the protein is "grown." A specific three-term sequence of bases in the mRNA, called a codon, calls up a tRNA carrying the specific amino acid that is to be the next unit in the growing protein chain. For example, the codon cytosine. uracil. guanine is **translated** as "leucine."

The existence of 64 – 20 = 44 excess codons allows a valuable redundancy in the genetic code. It also permits the signaling for the start and end of the protein chain.

You Need to Know

- Fisher projections
- D and L stereochemical designations
- α and β anomers
- Nucleotides and nucleosides

Solved Problems

Problem 15.1 Write structural formulas for (i) an aldotriose, (ii) a ketohexose, (iii) a deoxypentose.

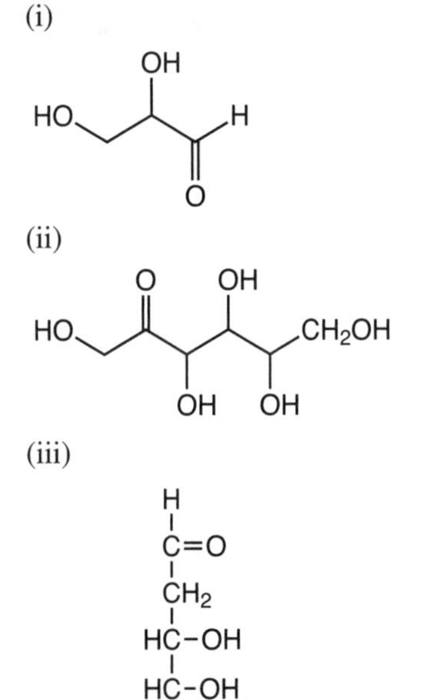

Problem 15.2 How do epimers and anomers differ?

Epimers differ in configuration about a single chiral center in molecules with more than one chiral center. Anomers are epimers in which the chiral site was formerly a carbonyl C.

Index